THE RHETORICAL MEDIATOR

THE RHETORICAL MEDIATOR

*Understanding Agency in Indigenous
Translation and Interpretation through
Indigenous Approaches to UX*

NORA K. RIVERA

UTAH STATE UNIVERSITY PRESS
Logan

© 2024 by University Press of Colorado

Published by Utah State University Press
An imprint of University Press of Colorado
1580 North Logan Street, Suite 660
PMB 39883
Denver, Colorado 80203-1942

All rights reserved
Printed in the United States of America

ASSOCIATION of UNIVERSITY PRESSES   The University Press of Colorado is a proud member of the Association of University Presses.

The University Press of Colorado is a cooperative publishing enterprise supported, in part, by Adams State University, Colorado State University, Fort Lewis College, Metropolitan State University of Denver, University of Alaska Fairbanks, University of Colorado, University of Denver, University of Northern Colorado, University of Wyoming, Utah State University, and Western Colorado University.

∞ This paper meets the requirements of the ANSI/NISO Z39.48-1992 (Permanence of Paper).

ISBN: 978-1-64642-529-7 (hardcover)
ISBN: 978-1-64642-530-3 (paperback)
ISBN: 978-1-64642-531-0 (ebook)
https://doi.org/10.7330/9781646425310

Library of Congress Cataloging-in-Publication Data

Names: Rivera, Nora K., author.
Title: Rhetorical mediator : understanding agency in Indigenous translation and interpretation through Indigenous approaches to UX / Nora K. Rivera.
Description: Logan : Utah State University Press, [2023] | Includes bibliographical references and index. | Text in English and Spanish.
Identifiers: LCCN 2023023521 (print) | LCCN 2023023522 (ebook) | ISBN 9781646425297 (hardcover) | ISBN 9781646425303 (paperback) | ISBN 9781646425310 (ebook)
Subjects: LCSH: Indigenous peoples—Languages—Translating—Research. | Communication of technical information—Research. | Translating and interpreting—Technological innovations—Research. | Linguistic rights—Research. | Indigenous peoples—Interviews. | Translators—Interviews.
Classification: LCC P120.I56 R58 2023 (print) | LCC P120.I56 (ebook) | DDC 418/.02—dc23/eng/20230719
LC record available at https://lccn.loc.gov/2023023521
LC ebook record available at https://lccn.loc.gov/2023023522

Partially funded by Chapman University.

Cover illustration by Nora K. Rivera

For Isabelle and Ed
y para ti abuelita en donde quiera que te encuentres

CONTENTS

List of Illustrations ix

Foreword: Doing the Work with Honesty, Care, and Respect
LAURA GONZALES xi

Preface xvii

Acknowledgments xxi

Introduction 3

1. Intersecting Theories and Disciplines 12
2. Designing the Research 27
3. Empathizing 41
4. Defining the Issues 64
5. Synthesizing Needs and Issues 94
6. Ideating and Re-Designing 129

Conclusion 147

Notes 153
References 157
Index 165
About the Author 177

ILLUSTRATIONS

FIGURES

1.1.	Literature informing this study	13
2.1.	Design thinking process by the Stanford d.school (2020)	31
2.2.	Design thinking process followed during the event	32
3.1.	Examples of individual user empathy maps	46
3.2.	Collective empathy map	47
4.1.	Examples of individual testimonio maps	71
4.2.	Collective testimonio map	72
5.1.	Motivations of Indigenous interpreters and translators	97
5.2.	Challenges of Indigenous interpreters and translators	99
5.3.	Feelings of Indigenous interpreters and translators	103
5.4.	Self-perception of the role of Indigenous interpreters and translators	104
5.5.	Needs identified by Indigenous interpreters and translators	106
5.6.	Issues identified by the participants in the roundtable	113
5.7.	Civic engagement activities in which participants take part	127
6.1.	Issues and solutions	130
6.2.	Vertical user interface (UI) of Indigenous interpreting events	131
6.3.	Peter Morville's (2014) user experience honeycomb	133
6.4.	User experience honeycomb with equity factor at its core	133
6.5.	Image shared on Twitter by ENES Morelia promoting the project *Traduciendo Juntas* (ENES 2019)	142
6.6.	Comic strip translated by Colectivo Uantakua (Luz 2019)	143
7.1.	The user interface of university classrooms	150

TABLES

2.1.	User empathy map adapted from Scott Wible (2020)	36
2.2.	Testimonio map	38
2.3.	Spanish-English glossary	40
3.1.	Luis's empathy map	48
3.2.	Mariana's empathy map	50
3.3.	Gabriela's empathy map	51

3.4.	Victoria's empathy map	52
3.5.	Pedro's empathy map	54
3.6.	Natalia's empathy map	55
3.7.	Claudia's empathy map	56
3.8.	Alejandro's empathy map	57
3.9.	Samuel's empathy map	58
3.10.	Abril's empathy map	59
3.11.	Lucas's empathy map	61
3.12.	Amanda's empathy map	62
4.1.	Rosa's testimonio map	74
4.2.	Carlos's testimonio map	76
4.3.	Alejandro's testimonio map	78
4.4.	Magdalena's testimonio map	81
4.5.	Antonia's testimonio map	84
4.6.	Claudia's testimonio map	85
4.7.	Lourdes's testimonio map	87
4.8.	Julia's testimonio map	89
4.9.	Valeria's testimonio map	91
5.1.	Baseline characteristics of participants who contributed to the interviews	96
5.2.	Baseline characteristics of participants who shared testimonios	112

FOREWORD

Doing the Work with Honesty, Care, and Respect

Laura Gonzales

I was recently part of an academic conversation where a panelist urged that scholars at all levels should work harder to "separate scholarship from advocacy." This claim is long-ingrained in Western knowledge systems—academic and otherwise. In academia, and in technical communication research contexts more specifically, there are ongoing and longstanding demands to remain "neutral and objective" in our research—demands that researchers provide "both sides" of clearly delineated arguments in order to be respected. Yet, as scholars such as Natasha Jones and Miriam Williams (2018) consistently demonstrate, technical communication, and research more broadly, is always imbued with value systems. Scholarship is always advocating for something. Indeed, much existing scholarship advocates for white supremacist values while ignoring and degrading nonwhite, non-Western perspectives (Haas, 2012).

In *The Rhetorical Mediator*, Nora K. Rivera shows us another way. Rather than shying away from naming scholarship as advocacy, Rivera positions this book as advocating for Indigenous language rights and liberation—and she invites technical communicators and user experience (UX) researchers to do the same. This book traces Rivera's collaboration with various communities, stakeholders, and participants, including the Centro Profesional Indígena de Asesoría, Defensa y Traducción, an NGO in Oaxaca de Juárez, Mexico, that advocates for Indigenous rights across Mexico, Latin America, and the world. Through this collaboration, Rivera demonstrates what it can look like for technical communication and user experience researchers to learn about and embrace Indigenous approaches to UX when we engage with community-based research in various contexts. Drawing from and extending design thinking frameworks, Rivera explains that if technical communication and user experience researchers truly want to work to redress oppression in their research, then we should not only acknowledge but intentionally join decolonial research practices already being enacted in various capacities by Indigenous people.

Structuring her book largely around the five phases of popular design thinking models (Empathizing, Defining the Issue[s], Ideating, Prototyping, and Testing), Rivera illustrates how user-experience researchers can build from important foundations for engaging participants and stakeholders while also asking important questions. For example, Rivera encourages design thinking researchers to consider: what does empathy mean through an Indigenous cosmovision that makes space for emotion and complexity rather than strictly aiming for efficiency? For Rivera, empathizing with participants is about much more than trying to "put yourself in someone else's shoes." Empathizing requires an attunement to testimonios—to narratives that "carry an underlying factor that urges civic engagement to produce social change." Rivera's connections between UX and testimonios is, in my opinion, one of the strongest contributions of this book. As Rivera explains, testimonios are about much more than venting—they are a process for collective listening, community building, and turning emotion into action. When technical communication and user experience researchers engage with frameworks for listening that expand beyond white, Western, monolingual ideologies, we can begin to better understand how methods of participation commonly used in our professional practices and methodologies can be exclusionary to marginalized communities. Asking someone to simply "provide feedback," "identify pain points," or "trace their journey" with an interface assumes that those individuals come from particular (i.e., white) traditions in which power and privilege don't influence participation.

When user experience researchers ask participants to share their perspectives, Rivera suggests we should consider how these perspectives are embodied and tied to lived experiences that may require processing and sharing. Through incorporating testimonios in their practices, UX researchers who understand, embrace, and respect Indigenous epistemologies can make space for acts of *desahogo* that can provide some release for people carrying a "distressful sentiment that keeps a person on the brink of not being able to breathe." Too often, UX research claims to seek participation and engagement while also prioritizing efficiency. For Indigenous communities and other communities of color, efficient models of participation may be extractive, superficial, and oppressive, seeking to assimilate or undermine "outlier" perspectives in the name of democratization and under the guise of equality rather than justice.

Rivera describes a two-day event created with and for Indigenous language interpreters, where participants and collaborators got the opportunity to discuss current issues impacting Indigenous communities, and Indigenous language interpreters and translators more specifically, all over the world. Using empathy mapping as a primary method, Rivera analyzes interviews that she conducted with Indigenous language interpreters and translators in this space, and she also provides insights from group conversations. Through this analysis, Rivera demonstrates the pitfalls of trying to simply follow design thinking protocols without providing space for participants to share their stories and build trust. Often, in UX research, we focus on individualistic perspectives gathered through different methods that help us identify larger patterns in the data. Contrastingly, the conversations that took place in Rivera's project foster collaborative sharing, healing, and strategizing. As Rivera explains, by engaging in design thinking that included opportunities for participants to share their testimonios, the group was able not only to make space for individual desahogos but also to build up from individual testimonios to a larger collective story and breaking point that led to change. As Rivera explains, through their conversations, Indigenous language interpreters and translators at this event were able to develop a "collective metatestimonio where the group built from one conversation to another until reaching a point of a collective desahogo that yielded the conscious feeling of 'enough is enough' of the group." This point of collective breaking and healing is a critical component for activities organized not just for or on behalf of but, more important, *by* and *with* marginalized communities. Without this space for sharing and processing, UX activities and other feedback and conversational practices may remain superficial and only to the benefit of those who do not share in the experiences of oppression being discussed or targeted.

An Indigenous approach to UX, as Rivera illustrates, is not about providing insights, detailing pain points, and testing prototypes for the benefit of corporate stakeholders. Instead, Indigenous approaches to UX provide space for building collective action toward the redressing of oppression for and with Indigenous people. In this way, Rivera's model for Indigenous UX directly addresses technical communication's social justice turn, which emphasizes moving from critique to action to directly redress oppressive structures and systems (Walton, Moore, and Jones 2019).

Rather than outlining the specific chapters in Rivera's book (which she does herself brilliantly in the preface and introduction), I conclude by sharing two specific strategies that Rivera executes and models for other technical communication and user experience researchers. First, Rivera makes clear that in order to work with community partners through a social justice perspective, researchers need to research how colonization impacts that community. In the case of Indigenous language translators and interpreters, Rivera highlights how Western notions of writing and documentation have been, and continue to be, imposed upon Indigenous communities. As Rivera clarifies, interpretation was always a part of communication for Indigenous people, for whom oral communication and storytelling are central methods of documentation. Yet, through colonization, "from a European lens, which regarded its own rhetorical and composition systems as objective and factual, Mesoamerican rhetorical traditions became known as unreliable and unstable practices." Thus, imposing methods of communication, research, and collaboration on Indigenous people can perpetuate colonial violence that erases Indigenous values. As Rivera demonstrates, Western alphabetic writing systems are used to dominate and erase Indigenous languages. At the same time, UX research methods that foreground Western modes of participation can also contribute to this oppression, even in projects framed as having social justice agendas. To embrace and practice social justice methods, Rivera argues, researchers need to understand the historical underpinnings of colonization and recognize how colonial ideologies are still at play.

When discussing community-based research, I've often been asked the questions, Who should be doing this work? How do we do work with communities we don't belong to? How do we contribute to social justice issues that don't directly impact us personally?

One of the most powerful contributions Rivera provides the fields of UX and technical communication is the careful, thorough, and continuous way in which she addresses her own positionality, "not as Indigenous woman but as a Mestiza." From the opening chapters tracing her work on the Mexico-US Borderland, to the way she attunes to positionality as a listener of testimonios, to the way she concludes the book by urging "Mestize, Latinx, and Hispanic scholars to grapple with and work through our own relationships with indigeneity in ethical ways," Rivera so clearly demonstrates that positionality is not a simple statement researchers make at the beginning of a project. Researchers shouldn't just name their whiteness and then move on from it. Instead, as Rivera shows us, researchers can be up front about why they are choosing to do work in a community, centralize the perspectives of people from that community, and make contributions to community advocacy work without centralizing their own needs, values, and perspectives. As a whole, Rivera's work shows us that there is a way to ethically collaborate across difference—if those of us with the most privilege are willing to read and listen beyond our own perspectives; put our personal agendas aside; and contribute to, rather than try to redirect or influence, the activist work already being enacted around our classrooms, our workspaces, and our institutions.

REFERENCES

Haas, A. M. 2012. "Race, Rhetoric, and Technology: A Case Study of Decolonial Technical Communication Theory, Methodology, and Pedagogy." *Journal of Business and Technical Communication* 26 (3): 277–310.

Jones, N. N., and M. F. Williams. 2018. "Technologies of Disenfranchisement: Literacy Tests and Black Voters in the US from 1890 to 1965." *Technical Communication* 65 (4): 371–386.

Walton, R., K. R. Moore, and N. N. Jones. 2019. *Technical Communication after the Social Justice Turn: Building Coalitions for Action*. New York: Routledge.

PREFACE

Dramatic changes in the demographic of immigrants arriving in the United States have brought to light the inadequate systems to address the needs of the hundreds of Indigenous language speakers seeking asylum at the US-Mexico border. Whereas Indigenous diasporas throughout the Americas have been ongoing since before European immigrants settled in these lands, violence and poverty have forced more and more Indigenous people to migrate to the United States during the last decades.

Before the year 2000, the great majority of undocumented immigrants came from Mexico, and approximately 90 percent were men who came to the United States to work. Among these statistics was my uncle, who moved to New Mexico as part of the Bracero Program, and my cousin, who joined the thousands of agricultural workers in the 1980s. My uncle died from cancer caused by pesticides, and my cousin died in a truck accident transporting farm workers without safety measures. My uncle and cousin are only two of many stories of Mexican immigrants who navigated inadequate systems that ignored their basic needs at the turn of the new millennium.

Between 2012 and 2019, however, the demographic of immigrants shifted drastically. After the year 2012, the percentage of arrests of undocumented immigrants from Honduras, Guatemala, and El Salvador increased consistently (O'Connor, Batalova, and Bolter 2019). By 2018 more than half of the undocumented immigrants that arrived at the US-Mexico border were from these three countries, and many of these immigrants were monolingual speakers of Indigenous languages (Jawetz and Shuchart 2019). Another major shift was that these undocumented immigrants were not primarily men anymore; whole families and unaccompanied minors made their way to the Borderland in large groups known as the caravans.

This Indigenous diaspora became more visible as a result of the shortage of Indigenous interpreters to meet the needs of the Indigenous immigrants that arrived at the Borderland. Indigenous interpreters became highly coveted in the United States but also became trapped in the middle of a dogmatic rhetoric of immigration emerging from the Trump administration between the years 2017 and 2019, dogmatic rhetoric that without a doubt would have continued had it not been for the major global disruption of the COVID-19 pandemic.

The caravans exposed the many systemic issues inside the US government in relation to immigration. Most importantly, this diaspora revealed an immigration system that only thinks of Latin Americans as Spanish speakers and is only prepared to "process" undocumented immigrants who speak Spanish. This issue, coupled with the lack of professionalization systems for Indigenous interpreters, plus the government policies of the week (literally, because they changed almost weekly), and the politicized rhetoric against immigrants, exacerbated a situation at the Borderland that caused hundreds of asylum seekers to live in tents in the Chamizal Park in Juarez, Mexico, and ultimately triggered a mass shooting at a Walmart in El Paso, Texas, a hate crime against the many Mexicans and Mexican Americans who work and frequently shop at this local store. The mass shooting took place on August 3, 2019. The first International Unconference for Indigenous Interpreters and Translators, the event at the core of this study, took place on August 8 and 9, 2019. This was the context in which my collaboration with Indigenous interpreters and translators took place.

Interactions between El Paso, Texas, and Juarez, Mexico, have been peaceful for as long as I can remember, albeit what the news and politicians often portray. Thousands of commuters live on one side and work on the other, and Mexican American students of all ages living in Juarez cross the border daily to attend schools in El Paso. Although the car lanes at the ports of entry are long, the vivid sounds of people crossing and street vendors talking and playing music made the tiresome commute entertaining before the Trump administration. To stop asylum seekers from crossing into the United States, the Trump administration overwhelmed the border with barbed wire fences and other "reinforcing measures" that produced jams at the ports of entry for up to four hours at a time, causing tremendous distress in the local community, disturbing the many transborder students who live in Juarez and (legally) go to school in El Paso, and, most importantly, provoking a traumatic strain on the asylum seekers stranded, living in tents, at the Chamizal Park in Juarez. As a Borderlander, I witnessed the consequences of every single issue expressed by the Indigenous interpreters and translators I met during the event at the center of this study. My journey into this research started at the end, at seeing firsthand the global implications of unstable governments, low and irregular wages, loose professionalization systems, lack of awareness about Indigenous matters, and, particularly, discrimination.

I finished writing the contents of this book in the midst of the COVID-19 global pandemic that forced our world to depend on digital technologies. The pandemic exposed many other issues of inequality in digital spaces. While some of us had the privilege to work from the safety of our homes with the help of technology, others had no choice but to continue business as usual with little to no protection against the virus. Race, as Anibal Quijano (2000) points out, continues to be "the fundamental criterion for the distribution of the world population into ranks, places, and roles in the new society's structure of power" (535). And despite the marked technological inequalities, Indigenous organizations have found in social media a powerful ally that the global pandemic also amplified. Indigenous social media advocacy has exploded, giving their fight global visibility, which I am sure will continue to expand in the future.

As I reflect on my participation as a collaborator, ally, and accomplice of Indigenous interpreters and translators, I think of my own journey navigating multicultural and multilingual spaces throughout my career as an educator. I think of the many students I have taught at US schools whose linguistic and cultural challenges are not much different from the issues Indigenous people face in Latin America. I think of the hundreds of Tarahumaras I used to see at the Bridge of the Americas as I crossed the border from Juarez to El Paso. I think of the mass shooting. And one image comes to mind, a small sculpture of children playing, holding hands, located at the Bridge of the Americas. The sculpture has an inscription in Spanish signed by Grupo Intercitadino that reads as follows:

> Soy parte de un círculo unido por amor y compañerismo, para llevar a cabo una misión de ejemplo, de ayuda mutua, y de progreso para la humanidad. Un ejemplo de amor en las razas, costumbres, e idiomas, una muestra de trabajo mutuo. —Grupo Intercitadino. Agosto 1999.
>
> [I am part of a circle united by love and partnership to carry out a mission of example, mutual help, and progress for humanity. An example of love among races, traditions, and languages, a demonstration of mutual work. —Intercity Group. August 1999.]

This study is a clear work of language activism that advocates for Indigenous language rights and Indigenous language practices, which Western scholarships and systems have greatly sidelined. Ultimately, I hope this work will guide those individuals in the legal, medical, and educational sectors who work with Indigenous individuals to consider the moral and ethical obligations that we all have not only to raise awareness about Indigenous language rights but also to enact upon these rights.

ACKNOWLEDGMENTS

This book was written in the US-Mexico Borderland, in spaces located in Santa Teresa, New Mexico; El Paso, Texas; and Juarez, Mexico, on the traditional lands of the Tampachoa, Mescalero Apache, Chiricahua Apache, Lipan Apache, Tigua, Piro, and Sumas. I hope that this land acknowledgment echoes throughout the book.

A significant portion of chapter 5, "Synthesizing Needs and Issues," was originally published in "Understanding Agency through Testimonios: An Indigenous Approach to UX Research," in *Technical Communication*, volume 69, number 4, in November 2022.

Muchísimas gracias a todas las compañeras y los compañeros de CEPIADET, a Tomás López Sarabia, Edith Matías Juan, Abigail Castellanos, Elena Ortega, Flavio Reginaldo Vásquez López, y a todas las y los intérpretes y traductores con los que trabajé en Oaxaca el verano de 2019. Mil gracias por compartir sus historias y sus conocimientos conmigo. Tlaskamati, Xquishepe', Sania, Guruare', Ki'imak óolal yéetel teés, Ndivé'e, Wokolawal, Koloval. El trabajo de cada una y cada uno de ustedes inspira mi trabajo. Your work inspires mine.

A huge thank you to Dr. Laura Gonzales, my mentor and friend. Since the day we met, your support has been instrumental throughout my academic journey. Your guidance gave me strength and confidence when I needed it the most. Another instrumental person has been Dr. Victor del Hierro. You and Laura are the reason I stayed in a doctoral program that sometimes made me doubt I belonged. Thank you both for inspiring me. Thank you, Dr. Lucía Durá. Your mentorship and support throughout the years have been key. I still remember our first meeting at the Starbucks by UTEP and how you convinced me to venture into a program I wasn't sure was for me. To Dr. Yolanda Chávez Leyva, many thanks for becoming part of my dissertation committee. I am a huge admirer of your work. As a Chicana historian who has worked with Indigenous communities, your feedback has been crucial to my research. A special thank you to my good friend Dr. Mónica Morales Good for inviting me to collaborate with Indigenous interpreters and translators. Colaborar contigo desde nuestros inicios en esta aventura en los cursos de maestría ha sido un privilegio.

Agradezco a mi mamá y a mi papá. Gracias por su cariño y sus enseñanzas. A mis hermanas Melina, Tania, Flor, y a mi hermano Helí. Gracias por su amor y apoyo. Melina, muchas gracias por tu apoyo incondicional sister. Thank you for being my rock, for always being there next to me en las buenas y en las malas (y gracias por ayudarme con la traducción de los términos legales).

Gratitude and love are not enough words to demonstrate my appreciation for the patience and support my daughter, Isabelle, and my husband, Ed, have given me throughout the years. As you said, Isabelle, you have never seen me not attending school. I raised you while attending school part-time and working full-time. None of this would have been possible without the unconditional support of my husband, who stepped up when school and work became tough. Remember when we met? Your Spanish was as bad as my English. So, Isabelle and Ed, this book is *our* achievement.

THE RHETORICAL MEDIATOR

Introduction

A BRIEF HISTORY OF INTERPRETATION AND TRANSLATION IN THE AMERICAS

Before the arrival of Europeans, Mesoamerican cultures maintained sophisticated systems to record stories, rituals, and histories. During the Postclassic period (from 900 CE to the European invasion), for example, Maya "writing and painting" thrived on the walls of their buildings and on "the pages of books" (Tedlock 1996, 23). Maya books were written using a complex system that combined phonograms (symbols that represent sounds) with logograms (symbols that represent ideas) that were *interpreted* only by elites (Houston, Robertson, and Stuart 2000). Evidence shows that *written* translation also took place as early as the first century between Nahua and Maya,[1] as seen in the Maya ceramic vessel found in Rio Azul, Guatemala, encircled with Nahua words written in Maya glyphs to describe the preparation of cacao (Macri 2005). Most of what we know today about this important writing system stems from the writings of the infamous Fray Diego de Landa, who, after burning the vast majority of the Maya codices in the Yucatán Peninsula, wrote a detailed description of their writing system (circa 1566) with the help of Maya interpreters and translators Juan Cocom and Gaspar Antonio Chi (Ceribelli 2013). The practice of writing with Maya classic symbols faded away gradually after the colonization of the Americas until its demise in the eighteenth century.

Like the Maya, the Mexica (a.k.a. Aztec) recorded their culture and history, *itolaca*, in pictographic books called *amoxtli*. The Mexica concept of writing, *tlacuilolitztli*, was intertwined with painting, so much that writing was also described as the action of using the red and the black ink, and writers or *tlacuilos* were trained in both painting and history (León Portilla 2010). Pictographic books (amoxtli) were stored in libraries called *amoxcalli*. Many amoxtli had an accordion-like structure, and they could be read linearly by spreading all the sheets at once, or they could be read in a nonlinear manner by strategically folding pages to apply a different interface flow of space and time, as it is evident in the *Codex Borgia*, one of the few surviving manuscripts written in the late fifteenth century (Díaz and Rodgers 1993). Mesoamerican books were read by *interpreters* trained in reading pictographs and glyphs and versed in the histories and rituals of their cultures.

When reading the stories in the images, interpreters had as much agency in constructing meaning as the images and as the writers (Rivera 2020). A clear illustration of this is the alphabetic version of the *Popol Vuh*, the book that tells the culture and the mythology of the K'iche' people. The original K'iche' authors of the alphabetic *Popol Vuh* quoted what the oral interpreters of the ancient hieroglyphic text "would say when they gave long performances, telling the full story that lay behind the charts, pictures, and plot outlines of the ancient book" (Tedlock 1996, 30); nevertheless, at one point in the book these *interpreters/translators* also become performers by "*speaking directly to us* as if we were members of a live audience rather than mere readers," as shown in the K'iche'-Spanish translation of the book by the friar Francisco Ximénez written between 1701 and 1703 (Tedlock 1996, 30). The agency of Mesoamerican interpreters became clear to the Spanish colonizers when they observed that the interpretation of a Mesoamerican book changed once the interpreter died or was replaced by a ruler (Mignolo 2003). From a European lens, which regarded its own rhetorical and composition systems as objective and factual, Mesoamerican rhetorical traditions became known as unreliable and unstable practices.

While Indigenous people continue to live under colonizing systems (Quijano 2000), long has it been since the cultures of the Americas first clashed with Europeans. Interpretation and translation practices have taken many forms since then. In central Mexico, tlacuilolitztli was replaced with alphabetic writing by training *Nahuatlatos*—a term given to Nahua speakers by Spanish colonizers—to write their Indigenous language using European alphabetic technology and then teaching them to translate Nahua into Spanish and vice versa (Alonso and Payás 2008). Spanish priests used Nahua as a lingua franca to make sense of the hundreds of languages spoken by the Indigenous communities across the New Spain before imposing Spanish. In spite of this, many of these American languages continue to function as official languages within Indigenous communities across the continent, particularly in Latin America, and Indigenous translators and interpreters continue to grapple with what it means to be mediators of languages, cultures, and worldviews.

UNDERSTANDING INDIGENOUS INTERPRETING AND TRANSLATING PRACTICES TODAY

Historiographies of Indigenous rhetorics and their influence on contemporary practices remain rare, abnormal subtopics of dominant Western academic traditions that persist in regarding Indigenous worldviews and practices as unreliable, especially in matters of technology and technical and professional communication. In places where Indigenous language translation and interpretation are greatly needed, Indigenous translators and interpreters face the lack of adequate systems to professionalize their field, withstanding public sector policies that do not align with the cosmovision of their cultures (Castellanos García et al. 2022).[2] They navigate monocultural and monolingual systems in multicultural, multilingual, and multiethnic societies.

Technical and professional communication—a field that often discusses matters of translation—and the field of translation and interpreting studies have not sufficiently examined the role of translators and interpreters within an Indigenous language context. This is surprising given that in Mexico alone there are 364 Indigenous linguistic variants treated as autonomous languages (INALI 2008) and that many monolingual speakers of these Indigenous languages migrate to the United States every year seeking a better, more stable life. These shortcomings prompted my research.

In 2018, I became involved in a collaborative community-based project to co-organize an event to collect resources to help in the professionalization efforts of Indigenous translators and interpreters. Under the leadership of the Centro Profesional Indígena de Asesoría, Defensa y Traducción (CEPIADET), an NGO from Oaxaca, Mexico, mainly composed of young Indigenous attorneys and interpreters, and with the help of scholars from the University of British Columbia in Canada, the Universidad de Veracruz in Mexico, and the University of Florida, we successfully co-produced the first International Unconference for Indigenous Interpreters and Translators in Oaxaca, Mexico,[3] converging approximately 370 participants from Mexico, Peru, and the United States, most of whom were Indigenous translators and interpreters.

This book analyzes the work carried out before, during, and after this event through an Indigenous approach to user experience research as a means to understand the role of agency within Indigenous translation and interpreting practices.

Drawing on the experiences shared by Indigenous interpreters and translators during the event, I primarily aim to examine how technical and professional communication (TPC), translation and interpreting studies (TIS), and user experience (UX) research can better support the needs of Indigenous language interpreters and translators. Specifically, this project is motivated by five guiding questions:

1. What are the needs of Indigenous interpreters and what are the critical issues they face?
2. How do Indigenous interpreters and translators understand and experience agency?
3. Why is it important to place equity rather than usability at the core of UX research?
4. How can analyzing Indigenous interpreting events as rhetorical negotiations of *truths* and recognizing the ambiguity within these negotiations help us understand agency in technical communication?
5. How can Indigenous approaches to UX help expand the fields of TPC and TIS?

This study advocates for Indigenous language practices that have been significantly sidelined by Western scholarship and systems. This work speaks directly to TPC scholars and UX researchers, urging them to include Indigenous technical communicators and their oral practices—interpreting, specifically—in disciplinary conversations. My work also speaks to TIS scholars, urging them to reexamine current translation and interpreting systems, for they are based on Western ideas and interfaces that disenfranchise Indigenous worldviews. Ultimately, this book calls upon those individuals working in the legal, medical, and educational sectors who work with Indigenous users to consider the moral and ethical obligations we all have not only to raise awareness about Indigenous language rights but also to enact upon these rights, and not in the way we think we should but in the way Indigenous people dictate.

FRAMEWORK AND DEFINITIONS

My work draws on Indigenous and decolonial theories and the scholarship of TPC and TIS. Because my primary research was conducted in collaboration with Indigenous people for Indigenous people, this work is strongly anchored in Indigenous theories with which I have a strong bond, not as an Indigenous woman but as a Mestiza who respects and values the common heritage and the shared history.[4] While Indigenous people widely use the term "Indigenous" to identify themselves and their communities, the term is complex and does not universally describe all individuals of Indigenous heritage. There are variants to this term depending on the region. In the United States, for example, it is common to use the terms "Native Americans," "Native American languages," and "Native Nations." In Mexico, the terms *pueblos indígenas* (Indigenous Peoples) and *lenguas indígenas* (Indigenous languages) are common. And in Peru, as I realized during the event in which I participated, it is more common to hear the terms *pueblos originarios* (Native Peoples) and *lenguas originarias* (native languages). Some members of Peru's Native Peoples also self-identify as *comuneras/os* (community members). In this research, I primarily use the term "Indigenous" because it was the term with which most of the participants in the event self-identified. Indigenous is also the term proposed by the United Nations (UN) Commission on Human Rights *Report of the Sub-Commission on Prevention of Discrimination and Protection of Minorities on Its 34th Session: Study of the Problem of Discrimination against Indigenous Populations* (a.k.a. Martínez Cobo Study) (1982) as the most generally accepted to refer to an individual who self-identifies as Indigenous and who is recognized and accepted by an Indigenous community. I also use the United Nation's working definition of the term "Indigenous community":

> Indigenous communities, peoples and nations are those which, having a historical continuity with pre-invasion and pre-colonial societies that develop on their territories, consider themselves distinct from the sectors of the societies now prevailing on those territories, or parts of them. They form at present non-dominant sectors of society and are determined to preserve, develop and transmit to future generations their ancestral territories, and their ethnic identity, as the basis of their continued existence as peoples, in accordance with their own cultural patterns, social institutions, and legal system. (UN 1982)

It should be noted that although there are similarities among Indigenous communities, many differences make each one of these communities unique, such as their cultures and languages. To be clear, Indigenous communities are multicultural, multiethnic, and multilingual.

Moreover, while some of the Indigenous theories discussed in my work are also decolonial, it is important to point out that there are significant differences between decolonial theories by Indigenous scholars and non-Indigenous scholars. The work by or with Indigenous people goes well beyond theoretical frameworks as Indigenous scholars contend for the "recognition of [their] sovereignty" and the recognition of their "immediate context" (Tuck and Yang 2012, 3). Although non-Indigenous scholars often develop decolonial theories, they are based on Indigenous epistemologies aimed at disrupting monocultural Western knowledge-making practices. Thus, decolonial thought is central for Latinx and Latin American scholars in and outside the United States. In spite of the different positionality of Indigenous and non-Indigenous scholars with Latin American heritage, our histories and cultures and personal and professional lives are all marked by the colonial moment, which triggered a historical record that ignored the histories and ways of knowing of people of color on the basis of race and ethnicity. Therefore, both Indigenous and non-Indigenous decolonial theories are important for analyzing Indigenous translation and interpreting practices.

Translation and interpreting (T&I) practices have not only been addressed in the field of TIS but have also been discussed in the field of TPC, hence the influence of these two fields on my work. On the one hand, TIS has challenged theories that imagine translators and interpreters as machine-like conduits with models that understand T&I events as mediated dialogical practices and with sociological perspectives that acknowledge the contexts of T&I events. More recently, decolonial views have also contributed to TIS by historicizing Indigenous T&I, highlighting power imbalances in T&I events when marginalized languages come into play, and problematizing the role of translators and interpreters' agency during T&I events. On the other hand, analyzing T&I from the perspective of TPC studies adds a critical perspective from which to see the role of agency in a T&I rhetorical event.

It is important to clarify that TIS scholars mark clear differences between translation and interpretation as professional practices (Angelelli 2004; Angelelli and Baer 2016; Biernacka 2008; Inghilleri 2012; Kleinert 2015, 2016; Niño Moral 2008; Strowe 2016; Tyulenev 2016; Wadensjö 2013), and thus this work also makes those distinctions. Whereas translation is seen as the act of interpreting *written* information to transfer it into a different written language, language interpretation is seen as the act of interpreting information *orally* from one language to another, or from oral to signed language and vice versa in the case of sign language interpreting.

Translation is an element of technical communication that has helped researchers and practitioners work toward more inclusive practices. TPC scholars have questioned neutrality in translation for decades, integrating contexts, power imbalances, and ethics in discipline conversation.

Some advocate for social justice approaches that address oppression in global arenas, particularly in the Global South (Agboka 2014; Savage and Agboka 2015),[5] and some incorporate human rights concepts to mold human-centered methodologies for technical communicators (Walton 2016).

TPC scholars have also advocated for participatory and localization research methods that can bring more just approaches to the theories and practices studied in this field (Dorpenyo 2020; Durá, Gonzales, and Solis 2019; Gonzales et al. 2022; Gonzales and Zantjer 2015). One such methodology is UX, a research approach widely used in technology fields but rarely applied to the contexts of Indigenous groups. UX is an interdisciplinary research methodology that focuses on the users and what they need and value with the purpose of designing better, more usable products and more effective content and processes. Users and usability are at the core of UX research. Peter Morville (2014), for instance, denotes that successful UX research should involve designing products, content, and processes that are useful, usable, desirable, findable, accessible, credible, and valuable. Whereas user-centered approaches are commonly used to examine the experiences of users in digital spaces and are starting to take hold in TIS through language-localization approaches—also primarily as they relate to translation in digital spaces[6]—these approaches are rarely used to examine the needs and values of Indigenous users, let alone to bring to light the complexities surrounding Indigenous translation and interpretation.

To analyze the experiences of Indigenous interpreters and translators, I examine the work done during the International Unconference for Indigenous Interpreters and Translators through *design thinking*, a solution-based approach aimed at solving complex problems through a process that typically involves the following phases as outlined by the Stanford d.school (2020): (1) Empathizing, (2) Defining the Issue(s), (3) Ideating, (4) Prototyping, and (5) Testing. Nevertheless, because my work is localized in the cosmovision of the Indigenous interpreters and translators who attended the event, it has important variations to the typical design thinking model, variations that I examine through a comparative analysis in chapter 2.

I analyze the Empathizing phase of the design process through interviews that are examined using user empathy maps. User empathy maps help initiate conversations focused on local contexts, eliciting new knowledge, which can disrupt preconceived positions (Wible 2020). Further, instead of defining the issues through personas (pseudo users that simulate real people's attitudes and behaviors), as in the typical design thinking process, I analyze this phase through *testimonios* by mapping the individual and collective *pain points* in each testimonio. Pain points are specific problems experienced by the users (Stanford d.school 2020). Testimonios are narratives that construct, and reconstruct (Mora 2007), a personal account that embodies a shared collective experience (Benmayor 2012). They carry an underlying factor that urges civic engagement to produce social change, hence their importance in my research. Many testimonios also involve the act of *desahogarse* (Rivera 2022b), the act of releasing a distressful sentiment that keeps a person on the brink of not being able to breathe (Flores and Garcia 2009). It is much more than just venting because desahogarse comes from experiencing extreme and painful sentiments. It is a cathartic act that openly releases suffocating anguish, providing a therapeutic feeling after the affliction is liberated from the body (Rivera, 2022b). While testimonios have been largely overlooked in UX research, they are a central part of this study.

Moreover, my work understands rhetoric as the way we negotiate truths (meaning) through the interfaces (relationships) we build between contexts, values, emotions, biases, power dynamics, loyalties, and dispositions. All in all, this book analyzes the data yielded by my UX research methodology through an Indigenous decolonial theoretical lens that weaves important scholarship from TIS and TPC as a means for analyzing the work of Indigenous translators and interpreters both as users and as technical communicators.

STRUCTURE OF THE BOOK

As previously stated, I examine the data gathered in my research through a design thinking process localized in the cosmovision of the Indigenous interpreters and translators who attended the event. Therefore, I chose to organize the chapters of this book around this process. Chapter 1, "Intersecting Theories and Disciplines," traces the intercultural and interdisciplinary scholarship that intersect the work that Indigenous interpreters and translators do day to day. The chapter highlights the significance of differentiating Indigenous and non-Indigenous decolonial theories. It also addresses the importance of examining Indigenous interpretation and translation not only from the standpoint of translation and interpreting studies but also from the technical and professional communication lens.

Chapter 2, "Designing the Research," examines my positionality as a Mestiza scholar collaborating with Indigenous groups through an autoethnographic approach that echoes throughout the book. This chapter introduces the design thinking process from which the methodology of the project is drawn and articulates interviews and testimonios as the central methods used in this UX study. In chapter 3, "Empathizing," I build on a critical approach to empathy to reflect on the importance of building long-term alliances with Indigenous communities and to preface the background of most Indigenous interpreters and translators as child language brokers. The chapter delineates my process for mapping interviews by displaying each of the user empathy maps I used to gather the raw data. Chapter 4, "Defining the Issues," puts forward testimonios as a UX method highlighting dialogue and desahogo as important Indigenous practices. This chapter also delineates my process when mapping testimonios and presents each of the testimonio maps I used to trace the experiences of the research participants.

In chapter 5, "Synthesizing Needs and Issues," I synthesize the raw data by identifying the participants' motivations, challenges, feelings, and self-perception of their role as Indigenous interpreters and translators. I also identify the participants' needs and the specific issues with which they grapple. Chapter 6, "Ideating and Re-Designing," emphasizes the importance of placing equity at the core of UX research. The chapter examines three projects presented by the participants at the event. It also identifies strategies to help improve employment conditions and opportunities for Indigenous interpreters and translators working in the public sector. The book concludes with the implications of this interdisciplinary project for practitioners, researchers, and educators.

The Rhetorical Mediator positions Indigenous interpreters and translators as technical communicators with rhetorical agency who understand the complexities of their work as acts of activism that help address the needs of their Indigenous communities. As the lack of awareness of Indigenous matters and discrimination continues to have a strong effect on Indigenous professionals, this book points out that TPC, TIS, and UX research can aid not only by raising awareness about Indigenous matters and practices but also by helping Indigenous professionals create methods and systems that better address their needs when working in the legal, medical, and educational fields. By and large this book is a work of language activism that advocates for Indigenous language rights and Indigenous language practices.

1
INTERSECTING THEORIES AND DISCIPLINES

A careful examination of the theories and the disciplines that intersect this work is crucial to understanding the work of Indigenous interpreters and translators as viewed from their own perspective and to unpacking areas of dissonance. While there are multiple academic disciplines that interact with the topics addressed in this study, my work draws from the scholarship of technical and professional communication (TPC) and translation and interpreting studies (TIS) and of both Indigenous and non-Indigenous decolonial theories, as shown in figure 1.1. Indigenous theories help situate this work in Indigenous ontologies and epistemologies. Tracing the connections and conflicts between Indigenous theories and decolonial work done by non-Indigenous scholars, Mestizes in particular, help identify areas of convergence from where to build long-term alliances. Moreover, examining Indigenous translation and interpretation from both TIS and TPC perspectives helps identify areas in which these two fields can interpolate to help in the professionalization of Indigenous interpreters and translators.

INDIGENOUS THEORIES

As previously stated, theories by and with Indigenous people differ from decolonial theories in several ways. One of the most important calls made by Indigenous theorists is their linguistic sovereignty demands, which not only advocate for the use of their own Indigenous languages but also call for validating orality as a legitimate technology, especially in relation to academic research. As Linda Tuhiwai Smith (2012) argues, "imperialism and colonialism brought complete disorder to colonized peoples, disconnecting them from their histories, their landscapes, their languages, their social relations and their own ways of thinking, feeling and interacting with the world" (29). Therefore, to understand Indigenous rightful demands of linguistic sovereignty, it is important to revisit Indigenous linguistic epistemological contexts through their

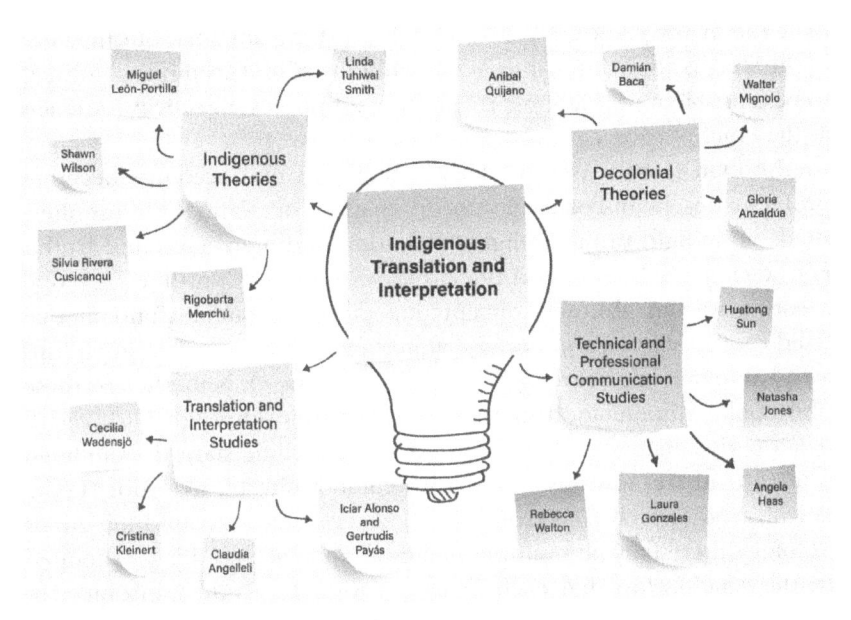

Figure 1.1. Literature informing this study

historiography. Nahua historian Miguel León-Portilla (2010), for example, sustains that ancient Mexicans maintained a sophisticated system to record history and culture called *itolaca*, loosely translated into "what is said about someone or something," in pictographic books (or amoxtli), which were *read* by narrating the stories and histories that the images symbolized (112). Although pictographic books were used as *written* historical and cultural records, they were orally interpreted by someone trained to do so, and each oral interpretation varied in its details.

Indigenous oral accounts, nonetheless, became meaningless to the Western conception of history because the Western European thought "situates the 'historical' only from what is [alphabetically] written, and thus legitimizes a colonial invasion as a heroic 'civilizing' mission" (Rivera Cusicanqui 1987, 9). Like the oral practices of ancient Mexicans, from an Aymara Indigenous cosmovision, the accuracy of an oral account is measured by the interests and perceptions of those who listen because Indigenous oral histories rely on testimonios and myths that function as interpretive mechanisms of historical recollections; *what* happens is not as important as *why* it happens (Rivera Cusicanqui 1987). An oral account changes each time someone tells it; "sometimes the change is simply in the voice of the storyteller. Sometimes the change is in the details. Sometimes in the order of events. Other times it's the dialogue or the response of the audience" (King 2003, 1), but the core of the story is always the same. Yet, the fluidity of oral accounts continues to be a problem in academic research that upholds objectivity and facts as the chief purpose.

Although from a Western lens factuality in oral accounts becomes problematic, León-Portilla's (2010) analysis of Nahua philosophers sheds some light on an Indigenous understanding of facts and truths. For ancient Nahua philosophers, *tlamatinime*, the truth of an oral account, was an abstract concept that meant to pursue fundamental fixed roots, *neltiliztli*, and because almost everything is temporary in this world, "no one might ever get to tell *the* truth" (255). Nonetheless, these philosophers understood that telling their stories through *huehuehtlahtolli*, or wise dialogues that consisted of storytelling narratives imbued with wise advice, was one way *a* truth could aspire to take root (León-Portilla 1991). Shawn Wilson (2008) asserts that "in an Indigenous ontology there may be multiple realities," and each reality is generated by the relationship that each of us has with the truth. "An object or thing is not as important as one's relationship to it," and therefore an Indigenous ontology is not constructed from "one definite reality" but from "different sets of relationships" (Wilson 2008, 73). Thus, Rivera Cusicanqui (1987) asserts, "We have no history, but histories, all of them of different depths" (9). Indigenous oral rhetorical practices have been addressed by Indigenous scholarship, particularly in their significance to storytelling. Nonetheless, they have yet to examine how Indigenous understandings of facts in oral accounts affect the Indigenous interpreting profession and how these views can help us better understand interpreting practices overall.

In relation to orality, Indigenous studies have also contributed with testimonios as a research method that affords rhetorical sovereignty (Lyons, 2000). Indigenous dialogues, like Nahua huehuehtlahtolli, gave birth to testimonios, narratives that construct, and reconstruct (Mora, 2007), a personal account—*a* truth—that embodies a shared collective experience. A testimonio "provides a structure within which events can be related and feelings expressed" (Smith 2012, 145). A testimonio is a narrative of "collective memory;" it is a method "for making sense of histories, of voices and representation, and of the political narrative of oppression" (145). Testimonios are critical for any research with and by Indigenous people because they make it possible for Indigenous stories to be told by their own voices, helping Indigenous individuals claim their right "to determine their own communicative needs and desires in the pursuit of self-determination" (Lyons 2000, 462). Testimonios have also given us insights into other important aspects of Indigenous ways of life, which are critical in academic research. For example, we know of the importance of community engagement for Indigenous people through the testimonio of Rigoberta Menchú (1984) in *I, Rigoberta Menchú: An Indian Woman in Guatemala*. In her testimonio, Menchú demonstrates the importance of communal affairs in her life with a narration of when at the age of ten her Maya community held a meeting to initiate her into adult life: "Then [my parents] made me repeat the promises my parents had made for me when I was born; when I was accepted into the community; when they said I belonged to the community and would have to serve it when I grew up" (49). Menchú's testimonio yields a glimpse at her community-based practices, which are critical to understanding Indigenous worldviews. For Indigenous individuals, the importance of "affirming connectedness" to a community relates to "issues of identity and place, to spiritual relationships and community well-being" (Smith 2012, 149–150). Although colonization disconnected many Indigenous people from their traditions, their "underlying beliefs are nevertheless carried on" (Wilson 2008, 94). Whereas testimonios and community-based approaches are widely used in Indigenous and decolonial research, they are not broadly accepted in mainstream research. Community-based research sporadically appears in technical communication scholarship, usually in research coming from the margins (Durá 2015; Rose et al. 2017; Singhal and Durá 2009), and testimonio is not a method used in UX research.

Conducting research with Indigenous people should include methods that are community driven because research—as previously exemplified by Menchú (1984)—needs to allow them to "fulfill [their] obligations or relationships to [their] community" (Wilson 2008, 110). Wilson (2008) suggests participatory action, talking circles, and personal narratives. Furthermore, Indigenous rhetorical sovereignty not only applies to the right to voice their own story but also to the ability to enunciate it in their own language. Indigenous communities call for the urgent need for language revitalization, and thus Indigenous scholars have proposed language-revitalizing projects as undertakings to support linguistic sovereignty (Driskill 2015; Smith 2012). Rivera Cusicanqui (1987)—a strong critic of decolonial discourse without action—proposes collective exercises of misalignment, like synthesized dialogues, in order to understand and delink both the investigator and the interlocutor from the places they both occupy in the colonial chain. Like Rivera Cusicanqui, many other Indigenous scholars have voiced their discontent with decolonial approaches that focus solely on theoretical frameworks—in their view, at the expense of Indigenous epistemologies—without carrying out any participatory research with or benefiting Indigenous communities (Driskill 2015; Itchuaqiyaq and Matheson 2021; Smith 2012; Tuck and Yang 2012; Wilson 2008).

Other scholars have put forward research that challenges Indigenous stereotypes, particularly in relation to technological abilities. It is crucial to utilize approaches that allow Indigenous people to represent themselves and to reframe inaccurate depictions of Indigenous people and communities because "there is a long Western rhetorical tradition of constructing American Indians—via print, visual, oral, and digital compositions—in stereotypical, essentialized, and fetishized ways that contribute to a larger, monolithic fiction of who/what is 'the American Indian'" (Haas 2015, 189). "The notion of 'origin'" often places Indigenous groups in an imaginary past that is "still, static, and archaic" (Rivera Cusicanqui 2010, 59). As explained by Angela Haas (2015), Indigenous people "are rarely represented, in fact, as contemporary peoples with complex identities and technological expertise" (189). The seminal work of Angela Haas (2007, 2012) problematizes Indigenous stereotypes concerning technical expertise by positioning technology not only as the tool that does the work but also as the work itself. Through this lens, technology includes, for example, the digital and analog tools used in translation and interpreting events as much as the written and oral languages used in such events.

DECOLONIAL THEORIES

Decolonial theories viewed through non-Indigenous positionalities are strongly influenced by postcolonial theories. To understand the differences between decolonial and postcolonial, we should recognize that these two terms came to the field of rhetoric and composition at different periods in time and from different geographic spaces (Rivera, 2020). Postcoloniality is a term that emerged from "political decisions following World War II, when much of South Asia and much of Africa were released from classical colonial control" (Villanueva 2016, v). In the late 1900s, the postcolonial option offered theorists like Gloria Anzaldúa and Cherríe Moraga the possibility of addressing issues of identity from the lens of the Other (Lunsford 1998). Scholars of the 1990s associated Anzaldúa's ideas with those of postcolonial theorists, such as Homi Bhabha, Gayatri Spivak, and Jacques Derrida, because they occupied the same academic space at the end of the twentieth century (Lunsford 1998). Decoloniality, conversely, sprouted right at the turn of the century from Latin American philosophers who claimed that while Latin American countries had gone through a process of independence, these States were not decolonized societies.[1]

Anibal Quijano (2000), in particular, argues that Latin America had rearticulated "the coloniality of power over new institutional bases" that continue to "maintain and reproduce racial social classification[s]" (567–568). Decolonial thought has been present in the Americas since the colonization of these lands and has always struggled to engage knowledge in a "paradigm of co-existence" (Mignolo 2005, 107). The "racially inferior" stratifications given to Indigenous people placed them in a "space of Otherness" in their own lands (Wynter 2003, 266, 300). The colonial moment also triggered complex intermixtures between ethnicities and cultures, complicating matters for those who believed in racial categorizations and pushing identity anxieties into the bodies of those of mixed blood, anxieties that continue to have an effect on people today. A "recognizable Other" continues to be viewed "as a *subject of a difference that is almost the same, but not quite*" (Bhabha 1984, 126, emphasis in original). Mestize scholars in Latin America and the United States have found in decolonial theories a link to Indigenous roots that challenge hegemonic epistemologies of power, language, and identity. And yet many decolonial theories by non-Indigenous scholars focus on rhetorical authority through metaphors, as Rivera Cusicanqui (2010) contends.

From her Mexican American Chicana positionality, Gloria Anzaldúa (2012), for example, proposed a new vision that enables Mestizes to "develop the agility to navigate and challenge monocultural and monolingual conceptions of social reality" (7). Adopting Indigenous knowledges as metaphors of resistance, she infused acceptance, tolerance, and pride into mixed identities of all sorts. She adapted *Nepantla* to signal her identity of in-betweenness—in between cultures, genders, and languages—and reconstructed the image of *Coatlicue* to embrace her English-Spanish forked tongue. She confronted an Anglo hegemony that views Mestizes in the United States with the same devalued lens through which Mestizes' ideologies sometimes view Indigenous people in Latin America because, as Quijano (2000) argues, Latin American systems are still modeled after Western European principles albeit being led by Mestizes. The colonial issues of power, race, ethnicity, land, and language of sixteenth-century Mexico, hence, found their way into the United States today, affecting the lives of Chicanxs, Indigenous people, African Americans, and Latinxs (Delgado, Romero, and Mignolo 2000). Although many decolonial scholars continue to adopt Indigenous epistemologies to frame theoretical work without actively engaging with

Indigenous communities, some have found significant ways to do both. Historian Yolanda Chávez Leyva (2003), for instance, has incorporated Indigenous concepts as frameworks to analyze Indigenous and Chicanx histories through teaching and learning experiences with Indigenous communities in Mexico. In a similar way, Gabriela Raquel Ríos (2015a; 2015b) has adopted Indigenous concepts as frameworks for participatory design and community engagement projects to highlight the rhetorical practices of farm workers and to build relationships while analyzing difficult conversations in the classroom. Nonetheless, there are still strong critics of decolonial scholarship coming from the Latinx academic community.

While decolonial theories grounded on Indigenous knowledge are widespread among Chicanx scholars, Eric Rodriguez and Everardo J. Cuevas (2017) argue that "toxic aspects of Chicanismo, such as problematic claims to land, practices, and knowledges" misappropriate knowledges "that were not ours to begin with" (232). Their claim, however, relies on a *translation* of José Vasconcelos's "Raza Cósmica" (Cosmic Race) and on what seems like a misinterpretation of the commentary provided by the translators as evidence of this misappropriation. In "Problematizing Mestizaje," Rodriguez and Cuevas (2017) claim that Vasconcelos "sees the resulting mestizo as more human" (231) although their source—which is not the original piece in Spanish but a translation—clearly states that Vasconcelos's text "was not meant as a prescriptive text submitting a form of eugenics. It was not a plan for improving the human species. Rather, it proposes an ethics that views complexity as an aesthetic value, and benevolence capable of producing a more dignified humanity" (O'Brien et al. 2013, 404).[2]

Further, Rodriguez and Cuevas (2017) argue that the nationalistic sentiment provoked by Vasconcelos's essay fuels an agenda that "could all at once *claim* Indigenous rights to settled land and *erase* specific understandings of Indigeneity and relations to land before colonization" (231).[3] While I do not negate the effects that Vasconcelos's glorification of Mestizes and his politics of homogeneity had and continues to have on Indigenous groups, I can only say that many Mexican and Mexican American scholars who conduct research with Indigenous groups do so out of admiration, respect, and cultural and historical synergy, and certainly not because of a land usurpation hidden agenda. Rodriguez and Cuevas's work places Chicanxs once more in a liminal space without any claims to Indigenous historical and cultural connections as if Mexicans and thus Mexican Americans and Chicanxs were once created out of thin air. Nevertheless, their work is important as it problematizes Mestize positionalities, which are unquestionably different from Indigenous contexts.

Non-Indigenous decolonial scholars also see in decolonial ideas a way to acknowledge the politics of language. Mignolo (1992), for example, has argued that after the colonization of the Americas, if a literary work did not conform with the mainstream canons of the European Renaissance, it was banished to the darker side. Consequently, pictographic writing was quickly transferred into alphabetic writing, and alphabets were created for some of the most commonly used Indigenous oral languages right after. "The Western Hemisphere," Damián Baca (2009) asserts, "is the only region on the planet where one writing system was so brutally and quickly imposed upon others" (232). This has caused major implications for the histories, languages, and epistemologies we privilege in composition and linguistic fields (Baca 2009; Mignolo 2009).

Similarly, postcolonial theorist Homi Bhabha (1994) questions the politics of the language of theory, deeming it "another power ploy of the culturally privileged Western elite to produce a discourse of the other" (20–21). Bhabha understands theoretical truth as a "negotiation" of culture and cultural concepts and not as a fixed idea that we can cage in one single universal category. This concept is not much different from the Nahua huehuehtlahtolli understanding of truth. Clearly, as Zairong Xiang (2016) states, applying decolonial frameworks to our academic work is for many of us not the act of appropriation that Rodriguez and Cuevas (2017) perceive but an act of *learning to learn* from Indigenous epistemologies.

TRANSLATION AND INTERPRETING STUDIES (TIS)

Decolonial theories have also intersected TIS in theoretical frameworks as scholars have used decolonial approaches to TIS to redress Indigenous historiographies. Scholars such as Icíar Alonso and Gertrudis Payás (2008) and Cristina Kleinert (2015), for example, have proved that the profession of translators and interpreters began in the Americas much earlier than Western scholarship asserts. As a means to communicate with colonized communities, the westernized professionalization of interpreters and translators began in the Americas during colonial Mexico at the turn of the sixteenth century (Alonso and Payás 2008; Kleinert 2015). Spaniards introduced the castilianized term *Nahuatlato* to refer to those who spoke Nahua and used this language as a lingua franca to create systems that facilitated the communication between

Spanish speakers and the multicultural and multilingual Indigenous groups in New Spain (Alonso and Payás 2008). Indigenous interpreters assumed roles localized in their geographical context. Whereas Nahuatlatos held official paid positions as language and culture mediators and used their skills as the first Indigenous people in the Americas to use alphabetic writing, to chronicle life in New Spain at the start of the sixteenth century, the Araucanian Indigenous interpreters held official paid positions that concentrated not only on language and culture mediation but also on diplomatic and military mediation to help mitigate the extreme political and military imbalances in Chile from 1605 to 1803 (Alonso and Payás 2008). In both cases, Indigenous translators and interpreters were seen as allies of the Spanish crown (Kleinert 2015). With the abolition of the Leyes de Indias after Mexico's independence from Spain at the start of the 1800s, however, the official paid positions of Indigenous interpreters were rescinded (Kleinert 2016). Clearly, the Indigenous languages of the Americas have always been engaged in language mediation, and translation and interpreting practices between European and Indigenous languages have existed since the colonization of the Americas.

In search of more standardized systems, furthermore, interpreting and translation scholarship in the 1970s advocated for a *conduit model* that followed Western views, supporting the scholarship of linguist and philosopher Michael Reddy in particular, which endorsed the idea of understanding translators and interpreters as neutral, invisible, machine-like conduits (Wadensjö 2013; Niño Moral 2008). Nonetheless, postcolonial and decolonial theories challenged these views. Homi Bhabha (1994), for instance, placed cultural diversity as a cultural difference that is "'knowledgeable,' authoritative, adequate to the construction of systems of cultural identification" (34). From his viewpoint, the process of translation is another colonial site of political and cultural friction and, therefore, translation is a political act of negotiating truths. Daniel Inclán Solís (2016) understands translation not only as a relocation of content but as a "process of emancipation" that one identity makes toward another through an opening produced by a critical view (57). Inclán Solís, like Bhabha (1994), positions translation as both a cognitive practice and a political practice that amplifies the voices of the marginalized. Nonetheless, and drawing on Rivera Cusicanqui's (2010) ideas, Inclán Solís (2016) cautions against a decolonial theoretical discourse in TIS that does not contribute to the well-being of Indigenous people, a discourse that instead has created another academic canon that speaks for the subaltern as ventriloquists do.

Other significant contributions to TIS from the standpoint of decolonial frames address the imbalances between Western and Indigenous cosmovision. For instance, Zairong Xiang (2016) claims that this inequivalence is particularly problematic when transferring content from an Indigenous mode or language to a Western mode or language. Similarly, Arturo Arias (2016) positions translation as "an instrument for disfiguring" because "there is always discrepancy"; therefore, translation—and thus interpretation—is "necessarily incomplete" (91–92). Furthermore, Cristina Kleinert (2015; 2016) centers her work on the need to refine and extend interpreting professionalization systems that address the specific needs of Indigenous interpreters in Mexico. Her important contributions address problematic concepts like agency (Angelelli 2004) and power (Quijano 2000) and relocate them to contexts surrounding Indigenous interpreters. She advocates for accredited professionalization systems as a critical factor for Indigenous interpreters to acquire agency, because historically, as asserted by Alonso and Payás (2008), there is a more institutionalized effort to language mediation when there is more quasi-equality among the power of the parties and cultures involved in the language negotiation (e.g., English-French, Spanish-English, Russian-Spanish, and Nahua-Spanish interpretation during colonial Mexico). Kleinert (2016), like Dalila Niño Moral (2008) argued before in different T&I contexts, also emphasizes the emotional maturity that Indigenous interpreters must have to withstand conflicting interpreting situations. Her fieldwork corroborates the co-constructing practices in which Indigenous interpreters take part when interpreting, dismantling theories that see interpreters as invisible conduits regurgitating information. Although few decolonial theorists working in TIS conduct research with Indigenous communities, their contributions are critical because they advocate for the recognition of Indigenous worldviews in this field.

Translation and interpretation studies have also borrowed theories from the field of sociology in an effort to prove the visibility of translators and interpreters. Theories such as Niklas Luhmann's (1996), which sees translation and interpretation as selection processes, Bruno Latour's (2005) actor-network theory, which helps us understand translators and interpreters as actors who can even out imbalances in a network—and thus in a translation and interpreting act—and Anthony Giddens's (1984) theory of action, which argues that agency is not intentional, have all influenced TIS (Tyulenev 2016). Claudia Angelelli's (2004) T&I work is also influenced by sociology. Her sociological perspective conceives interpreting as a co-constructing practice that draws on the inherent social values of the interpreter's families and communities, a theory that also refutes seeing T&I events as conduit processes. From her view, however, interpreters can still be impartial because their practice is situated in their professional contexts, and interpreters must follow "professional regulations, standardization, and certification" (98). Most important, Angelelli's studies show that some professional contexts make interpreters more visible, like the health industry, while other contexts make them more invisible, like courts and conference settings. In terms of the professionalization of interpreters, Angelelli argues that codes of ethics are based on limited understandings of what is right and what is wrong and that the focus of professionalization "should be broadened from terminology mastery and information processing to include other issues that are important in the interpreted communicative encounter, such as an awareness of the power that interpreters possess in the cross-cultural/linguistic encounter" (89). Although linguistic and cognitive skills are essential, these are partial skills needed to perform the role of interpreters, and a full range of abilities should be addressed when teaching interpreters, including teaching professional knowledge (e.g., comparative law for court interpreters, basic anatomy for medical interpreters, and international organization skills for conference interpreters) and interpersonal skills (Angelelli 2004; Angelelli and Baer 2016). Therefore, the professionalization of interpreters, and thus translators, should engage conversations between and across different fields in order to revitalize ontological and epistemological paradigms in TIS.

Other important contributions to TIS acknowledge the fluidity of language, problematizing ideologies of neutrality in both translation and interpreting practices. Cecilia Wadensjö (2013), for instance, sees interpreting as a mediated dialogical practice. She grounds her ideas on Mikhail Bakhtin's dialogic philosophy, which understands the meanings conveyed by language as "co-constructed between speaker and hearer(s)" (Wadensjö 2013, 41). She questions neutrality because, as she explains, even a formal and stoic poise on behalf of the interpreter can be perceived as a position that favors the public institution. Although ethical codes reject verbal activity outside what the interlocutors say, "in practice . . . these situations do indeed occur" (Wadensjö 2013, 191). Like Wadensjö, Moira Inghilleri (2012) defines translation and interpreting as instruments used "to reveal and represent the 'living dialogue' that contributes to the formation of worldviews, opinions, values, and beliefs that are formed and transformed by human interaction and intervention of different environments" (130). She acknowledges translation and interpretation as practices that *form* and *transform*, positioning the ever-present issue of impartiality as a paradox that in practice might not happen, particularly when professional "loyalty confront[s] questions of justice and individual conscience" (17). Furthermore, and drawing on the work of Angelelli (2004) and Inghilleri (2012), Anna Strowe (2016) asserts that because T&I events are shaped by asymmetries of power and because T&I professionals can act as advocates or as institutional gatekeepers, the education of translators and interpreters must include the subject of ethics. Evidently, translation and interpreting professionals not only mediate language and culture but also power, values, and even loyalties.

Although it is clear that interpreters mediate a lot more than language, many T&I programs and scholars, and many interpreters themselves, still adhere to outdated beliefs of neutrality and invisibility. For example, Agnieszka Biernacka's (2008) research with interpreters in Poland shows that while most interpreters clarify cultural, linguistic, and geographic differences, very few believe that these elements play an active role in the communication act. Similarly, Dalila Niño Moral's (2008) study in Alicante, Spain, shows that interpreters feel under substantial emotional tension as they navigate through the different contexts of their work while controlling the strong emotions triggered by the tragedies they witness because they feel that they are expected to *only* mediate language. Niño Moral's work sheds light on the different perception about the most important skills interpreters should have. In Niño Moral's study, service providers that contract interpreters believe that the most important skill is good character, which in their view consists of a combination of patience, empathy, respect, and resilience, whereas interpreting students believe that their most valuable asset is their ability to interpret, proving that interpreters' perceptions of their own practice is strongly influenced by the neutrality ideologies learned in their professionalization programs. As seen in this section, translation and interpreting studies, although grounded on Western ontology and epistemology, are shaped by multidisciplinary, multicultural, and multilingual practices that problematize the objectivity of the communication mediated between languages, cultures, disciplines, and professions. Still, TIS has a limited scholarship that examines T&I events from non-Western ontological and epistemological views, and in the case of TIS in the Americas, it is especially important to analyze translation and interpretation from the lens of Indigenous worldviews.

TECHNICAL AND PROFESSIONAL COMMUNICATION (TPC) STUDIES

Technical and professional communication studies has also expanded the conversations surrounding translation practices. An illustration of this is Jennifer Daryl Slack, David James Miller, and Jeffrey Doak (1993), who have argued that communication is always asymmetric, and thus there is always an ongoing struggle for power. Technical communicators, as translators, can never be transparent because they retrieve, reconstruct, and contribute to the meaning of language; therefore, technical communicators not only have power but also ethical responsibilities to ground their work "in an understanding of how power works" (Slack, Miller, and Doak 1993, 24). Moreover, Laura Gonzales (2018) argues that technical communication must recognize the practices of translators in a broad rhetorical context that connects multilingualism and multimodality because "translators have to manipulate and coordinate multiple modes simultaneously," positioning translation not as a neutral exercise but as a culturally situated practice that is cyclical and creative (60). Translation is not a simple act of language transmission, for it "require[s] extensive rhetorical negotiation" (Gonzales 2018, 21). Further, Yusaku Yajima and Satoshi Toyosaki (2015) argue that while most of us recognize that intercultural and interlingual communication construct and negotiate identities, "translation cannot be *fully* socially just" because translators, and thus interpreters, mediate between speakers of hegemonic languages (101). Yajima and Toyosaki, therefore, claim that we should make global language hegemony visible to engage in more ethical language negotiations. Yet, as Brent C. Sleasman (2015) notes, because most public systems are framed in colonial structures, most government agencies and government-funded organizations continue to support the conduit T&I model, undermining the practices of minoritized cultures—and oral practices in particular.

Moreover, social justice as a concept that challenges global inequalities has deeply influenced TPC scholars. Take for instance the work of Iris Marion Young (1990) on structural oppression. Young identifies the patronizing and punitive policies of bureaucratic institutions as structural oppression, adding that the way in which we conceptualize differences among groups and the way in which society marginalizes nonprofessionals in important decision-making processes perpetuates oppression. From this perspective, TPC is neither neutral nor objective; it is political (Jones 2016). To challenge TPC's innate imbalances, scholars have worked on reconceptualizing the field "to incorporate contexts of social justice and human rights" (Jones 2016, 3). Natasha Jones calls for TPC scholars to reflect on their own positionalities and enactments of power and to directly engage and take action on issues of social justice, inequality, and dehumanizing forces. There is a great need to openly discuss race and ethnicity in TPC scholarship through *antenarratives*, or reinterpretations of the past, to disrupt dominant narratives that focus on the effectiveness of technical and professional communication, negatively affecting marginalized groups (Jones, Moore, and Walton 2016). As Rebecca Walton (2016) argues, if we are to prioritize people, not only as users but as humans, then human dignity and human rights should be the first principle of human-centered design.

Translation has helped researchers and practitioners in technical communication work toward social justice. Global South perspectives within TPC in particular have contributed critical scholarship to social justice conversations that are informed by translation. TPC has the responsibility to promote practices that "project and advance issues about populations within the Global South, and provide resources (e.g., theories, methods, and cases) for addressing the challenges raised by research in and about the Global South" (Savage and Agboka 2015, 11). Because "research is always cultural" (Agboka 2014, 299), it is important for TPC to address the power dynamics inherent to cross-cultural global research. This book complicates and extends the important work that TPC scholars have carried out with Indigenous groups in the Americas (Gonzales 2021; Gonzales et al. 2022; Haas 2012; Itchuaqiyaq and Matheson 2021; Rivera 2022b). If we are to engage in social justice approaches, it is absolutely critical to examine contexts meaningful to the geographical spaces we inhabit through non-Western views.

In an interdisciplinary field like TPC, scholars advocate for participatory research that is action-driven, community-based, and user-centered in order to directly engage in issues of inequality (Jones 2016; Jones, Moore, and Walton 2016; Walton 2016). One example is user experience (UX) research. UX research is a user-centered methodology commonly used in TPC studies because it organically accommodates research driven by users' needs to solve technology-related problems. Redish and Barnum (2011) argue that TPC scholars primarily utilize UX research in content development. Given that technical communicators often work in multicultural and multilingual contexts, some TPC scholars advocate for using UX research to examine technical communicators as translators. For example, Suojanen, Koskinen, and Tuominen (2015) propose a view of the technical and professional communicator as the user's advocate and thus encourage TPC to include usability methods in translation pedagogy through a user-centered translation (UCT) approach aimed at enhancing the professional skills of translators.

Since its inception, user experience research has had strong ties to technical communication (Redish and Barnum 2011). Its focus on the user particularly appeals to scholars who work with frameworks that study usability in local contexts. Huatong Sun (2006), for instance, developed a cultural usability framework to argue that each user has "a local purpose and a social motive," and thus researchers should use localization to help repair issues caused by contextual misunderstandings (Sun 2006, 473). Gonzales and Zantjer (2015) propose that technical communicators who work in international contexts make use of culturally localized translation to better address the needs of global users, highlighting the various rhetorical strategies that multilingual speakers use to "adapt information in international contexts" (272). Studying the rhetorical strategies of multilinguals helps technical communicators create "user-centered global content" that localizes uses of language in the contexts, cultures, and experiences of international users (Gonzales and Zantjer 2015, 281). While the works of TPC scholars working with UX approaches have addressed important issues affecting technical communicators as translators, the word "technical" or "technological" rarely intersects with the term "Indigenous" in UX (First Nations n.d.; Rivera 2022b), and UX research that considers the needs of Indigenous technical communicators as translators and interpreters is scarce.

Indigenous communities in the Americas have always been multicultural and multilingual. Their complex and unique translation and interpreting practices have existed all along. Considering both TIS and TPC theories in this study is key because Indigenous translators and interpreters in their role as technical communicators work in judicial, medical, educational, and governmental environments that require them to translate and interpret highly technical information from one cosmovision to another. Analyzing these practices in the contexts of two different fields through a non-Western lens helps us recognize the significance of considering Indigenous approaches in UX research. In the next chapter, I describe how an adaptation of the traditional UX methodology used in technical fields can more adequately address the needs of Indigenous users. The next chapter also discusses the importance of reflecting on my own positionality as a Mestiza researcher working with Indigenous groups.

2
DESIGNING THE RESEARCH

POSITIONALITY

Indigenous communities in Mexico and Latin America have a relationship toward Western systems akin to Native communities in the United States and Canada. Yet, each context is different. One of the main differences is the greater number of monolingual speakers of Indigenous languages in Latin American countries, hence the higher need for Indigenous interpreters and translators. Another difference is the dominant group. Whereas systems in the United States and Canada are led by Anglos, Latin America is led primarily by Mestizes, a term first introduced by Spaniards in colonial Mexico to categorize individuals of Spanish and Indigenous mixed race. Today, the term describes individuals of Latin American background with mixed European and Indigenous ancestry who "favor the Western traditions and practices inherited from the Spanish colonial systems (e.g., public schools, public health institutions, and courts)" (Rivera 2022b, 10). Indigenous individuals also use the term "Ladino" or "Ladina" as a synonym for Mestize.

I am a Mestiza born and raised in Juarez, Mexico, a city of approximately 1.5 million people that borders the city of El Paso, Texas. My mom is from the small town of Pascual Orozco, located in the Sierra Madre mountains of Chihuahua, Mexico, neighboring a Mennonite settlement and Tarahumara communities. My dad is from the town of Matalotes, Durango, Mexico, a community of fewer than one hundred people also located near Tarahumara communities. As a Mexican from *La Frontera* (The Frontier), as Mexicans call the Borderland, I have always been aware of my double positionality: privileged when contrasted with the hundreds of Tarahumaras living in Juarez but not so privileged when compared with Anglos in El Paso. However, it was not until I began working with Indigenous organizations that I truly understood the privileges I am afforded as a Mestiza. Mexican Mestizes do not always draw attention to the positionality of our identities because we tend to boast about the kaleidoscopic distinctiveness of our mixed race. This seemingly inconsequential attitude toward race and ethnicity

https://doi.org/10.7330/9781646425310.c002

hides a systemic problem that places value on the color of our skin and the languages we speak. Some take pride in our European heritage, oblivious to the violent circumstances in which most of this legacy took place. Some take pride in our *ancient* Indigenous roots, romanticizing the mighty Aztec and undermining the many connections we share with our Indigenous sisters and brothers *today*. For better and for worse, I have done all of the above at different points in my life. I've been the Mexican who boasts about her European heritage, the one who takes pride in ancient Indigenous roots, and the one who sees and feels strong connections with Indigenous people today.

As a literal result of colonization, Mestizes cannot always pin down with precision our racial identity even if we conceive our historical roots in the Americas and not in Europe. This sting of knowing and not knowing is one of the effects of colonialism on Mestizes today. My dad, for instance, grew up next to Tarahumara communities, and although his own family practiced Indigenous traditions like agriculture and midwifery, he does not self-identify as an Indigenous person. My mom grew up in a place that is also next to Tarahumara communities and near Mennonite settlements. Like many people around that area, my mom's family strongly resembles Mennonites, but no one in her family self-identifies as one. Race is something that most Mexican Mestizes discuss on the bodies of others but hardly examine on our own bodies. In practice, the word "Mestize" is hardly used by Mestizes in Mexico. Nonetheless, we learn to self-identify for others in the United States who see us through the same lens we see Indigenous people in Mexico. US institutions, for example, do not see much difference between an Indigenous person from Mexico and a Mexican Mestize and a Mexican American, or between a Mexican and a Puerto Rican, or between a Colombian and a Peruvian; "We are all lumped in the same category: Hispanics, whether we speak Spanish or not" (Rivera 2022b, 10).

Indigenous-Mestize interactions are complicated as they have historically been imbued with power dynamics, contending loyalties, and discrimination. In academia, for example, Mestize scholars are often criticized for both working and not working with Indigenous communities. We are criticized for not working with Indigenous communities because it makes us look as if we negate our Indigenous roots, and we are criticized for working with Indigenous groups because it makes us look as if we are claiming an Indigenous ancestry to which we have no right, since we neither live in Indigenous communities nor speak their languages. Yet, we cannot absurdly think of Mestizes as Europeans because at one point in our history we were *born into* Western systems—violently. It is almost as absurd as saying that Chicanxs can't claim Mexican heritage because they were not born in Mexico or because they do not speak Spanish.[1] As complicated as it sounds, Mestizes share with Indigenous people history, a colonial wound, and many practices that have made their way through our cultural mixture.

Indeed, academic interactions between Mestizes and Indigenous groups are complex and not always ethical, provoking an understandable skepticism and mistrust among Indigenous groups. In spite of and because of these complexities, many Mestize scholars work with Indigenous groups in Mexico and Latin America because we value our commonalities, recognize our differences, and acknowledge the shared responsibility to work together.

I am a Mestiza scholar, and this study was conducted in collaboration with Indigenous people with the utmost respect and admiration for their knowledges, from the positionality of a Mestiza who values the common heritage and the shared history.

METHODOLOGY

This study, albeit vaguely addressing translation, is strongly slanted toward the investigation of Indigenous interpretation. As explained before, most academics understand translation as a written act and interpretation as an oral act, and Indigenous languages are rooted in oral practices. Although many Indigenous languages have alphabetic writing today, some still rely mainly on oral and performative practices. For this reason, Indigenous (oral) interpretation is a ubiquitous practice in places with large Indigenous communities, while Indigenous (written) translation is still very limited. It is also important to clarify that this work mainly addresses interpreting practices of Indigenous languages of Latin American territories because today most Native Americans in the United States and Canada are bilingual or monolingual speakers of English.

My methodology analyzes the work carried out during the International Unconference for Indigenous Interpreters and Translators in Oaxaca, Mexico, on August 8 and 9, 2019. The organizing team worked under the leadership of the Centro Profesional Indígena de Asesoría, Defensa y Traducción (CEPIADET) in collaboration with scholars from the University of British Columbia in Canada, the Universidad de Veracruz in Mexico, the University of Florida, and the University of Texas at El Paso. The project also gained input from the Mixteco Indígena Community Organizing Project (MICOP) in California, the Universidad Peruana de Ciencias Aplicadas (UPC), and the Biblioteca de Investigación Juan de Córdova. My involvement in this project was partially funded by a fellowship that I earned from the Kapor Center in Oakland, California, and the coalition of Women of Color in Computing at Arizona State University. This initiative intended to increase the representation of women of color in computing and related fields by providing guided mentorship in technical and professional communication and in user experience research.

As a co-organizer, I analyze this event through an Indigenous approach to UX research as a means of understanding the role of agency within Indigenous translation and interpreting practices. My work is framed by an Indigenous UX approach to translation and interpretation that (1) places equity, anchored in human rights rather than usability, at the core of UX research; (2) supports community-driven UX practices; (3) understands Indigenous interpreting events as rhetorical negotiations of *truths* and recognizes the ambiguity within these negotiations; (4) analyzes the cultural, professional, sociopolitical, and linguistic contextual spaces in which Indigenous interpreting events occur as user interfaces (UI); and (5) problematizes the role of agency in Indigenous translation and interpreting events.

Design Thinking as a Community-Driven UX Practice

To produce the International Unconference for Indigenous Interpreters and Translators, our team collaborated via videoconferencing from three different countries; Canada, the United States, and Mexico. We sought out to create a community engagement project that was action-driven, user-centered, and highly localized. As highlighted by Indigenous scholars, community practices are at the center of Indigenous ways of life (Menchú 1984; Rivera Cusicanqui 1987, 2010; Smith, 2012; Wilson 2008). Therefore, an *unconference* structure, where participants guide the discussions by becoming presenters, became the ideal collective approach to work with Indigenous interpreters and translators.

While the event was not engineered to follow a design thinking process, it organically moved in that direction. The event followed a design thinking structure similar to that proposed by the Stanford d.school (2020), shown in figure 2.1, which typically adheres to a process of five stages: (1) empathizing through interviews or shadowing; (2) defining the issue(s) through the creation of personas, story mapping, role objectives, decisions, challenges, or *pain points*—specific problems experienced by the users; (3) ideating through idea sharing, diverge/converge—a strategy where users first generate ideas independently (diverge) and then analyze them collectively (converge)—and "yes

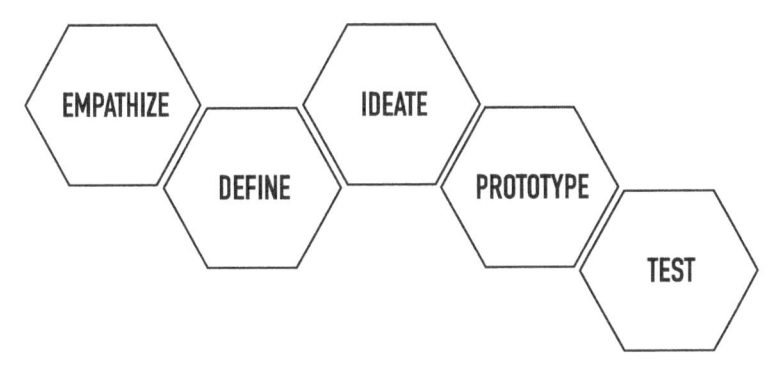

Figure 2.1. Design thinking process by the Stanford d.school (2020)

and . . ." thinking; (4) prototyping through mockups and storyboards; and (5) testing through role-playing (see figure 2.1). This study, however, analyzes the event's structure through an adaptation of the design thinking process proposed by the Stanford d.school (2020).

Our event lasted two full days and consisted of one large opening session, four major roundtable discussions, and one large closing session. The opening and closing sessions were open to the general public, and the roundtable discussions were restricted to participants who submitted applications and presentations ahead of time. It should be noted that although the great majority of the 370 participants from Peru, Mexico, and the United States were Indigenous interpreters and/or translators, the event also included the participation of academics and public officials, such as the magistrate of the Supreme Court of Puno, Peru, Dr. Hernán Layme Yépez; Luis Arturo Fuentes Gómez, assistant director of education and training of the Instituto Nacional de Lenguas Indígenas (INALI) of Mexico; and the Mexican federal judge, José Luis Evaristo Villegas.

CEPIADET wanted to ensure that the event included the diverse perspectives of various Indigenous groups as well as of various public institutions. Importantly, because many of the Indigenous languages from Mexico and Peru mentioned in this study have made their way to the United States through migration, this study concerns all three countries. Latin Americans and Caribbean-born individuals are the largest combined immigrant group in the United States and the fastest-growing demographic group (Vespa, Medina, and Armstrong 2020). Alison Cardinal (2022) argues that global migration has increased the complexities associated with cultural identities, exposing an in-flux "diversification of diversities" phenomenon known as superdiversity (2). Whereas superdiversities have always been present in the Americas,

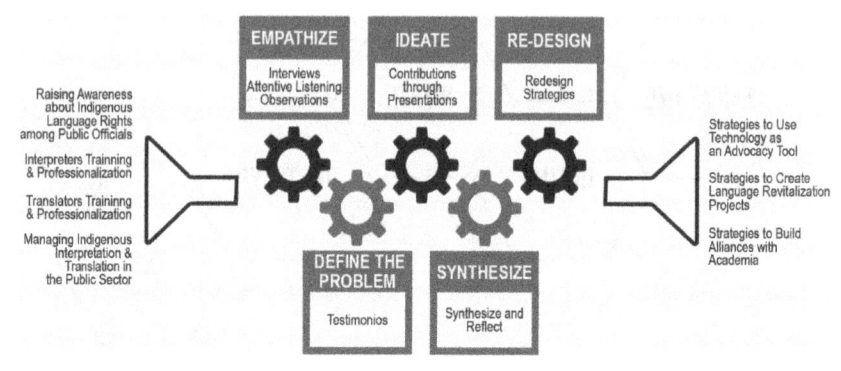

Figure 2.2. Design thinking process followed during the event

they have not always been visible. As a methodology that provides ways to untangle complex problems by examining issues through diverse perspectives, design thinking was ideal for this study.

As in the case of traditional conferences, the selected participants were placed in a roundtable discussion based on the topic of their presentation proposals, as envisioned by CEPIADET. The four major roundtable themes from where participants built their contributions were: (1) raising awareness about Indigenous language rights among public officials; (2) training and professionalizing interpreters; (3) training and professionalizing translators; and (4) managing Indigenous translation and interpretation in the public sector.

This study draws on my participation as a co-organizer of the event and as the moderator of Roundtable #4, which addressed questions related to managing Indigenous translation and interpretation in the public sector. Twelve participants contributed to the conversations in Roundtable #4: two Indigenous college students who also work as court interpreters and intercultural mediators in Oaxaca; five Indigenous interpreters from various communities from the Mexican states of Oaxaca, Chihuahua, Morelos, and Mexico, one of whom had the role of transcribing key points of the conversations, and thus had limited participation; one Indigenous translator who conducts language revitalization work in her community in Michoacán, Mexico; two Indigenous translators, interpreters, and activists from California; one translation and interpreting studies scholar from Peru; and one academic from the University of Antwerp in Belgium who has conducted work with Indigenous interpreters and translators in Peru. I examine the event and the conversations in Roundtable #4 as an adapted design thinking process structured around the following five phases (see figure 2.2).

Empathizing

The Empathizing phase followed during the unconference was similar to that of a typical design thinking process. We began with participants introducing themselves by giving their names, the name of their community, and a brief description of their professional background. While each individual spoke, the rest of the participants attentively listened without interrupting. This step helped situate our conversations while fostering harmony and compassion among those listening. A similar process took place in the other three roundtable conversations. Meanwhile, participants from all four roundtables volunteered for individual interviews that were conducted throughout the morning. These interviews and the attentive listening and observations happening in each roundtable were the central elements of the Empathizing phase. The interviews are analyzed as a research method in chapter 3.

Defining the Problems

Whereas the typical Defining phase in design thinking workshops consists of creating personas, fictional characters, or role-playing in order to identify the issues in question, our Defining phase organically became an open session of testimonios. This phase began with informal introductions of each of the twelve participants. Then, each participant took turns to informally talk about their own experiences, the strategies they use, and the civic engagement in which they participate. These dialogues took place during the second part of the first day of the event and also during the first half of the second day. The dialogues spontaneously became testimonios that offered a clear view of the issues faced by their communities. Each individual offered personal accounts of their experiences as well as the collective experiences of their communities before presenting some of the strategies they have tried out to grapple with the issues they face. Even though not all participants proposed strategies, most of them contributed with recounts of their unique experiences, which revealed the collective issues of their communities. These testimonios became the central element of this phase. Although testimonios are not used in UX research, they can be a powerful tool to define the problem in a design thinking process because listening to a person who has experienced the problem in question is more valuable than imagining how a person would feel or react to a problem through personas or role playing. This method is explained in detail in the methods section that follows, and the testimonios of the participants are analyzed in chapter 4.

Ideating

A traditional design thinking process often includes diverging/converging sessions, where participants generate ideas independently (diverge) and then analyze them collectively (converge). In a way, this is the strategy we followed here because some participants had already tested strategies independently with their own communities and were now presenting them and analyzing them collectively during our discussions. Whereas in theory each phase seems to be tidy and self-contained, in practice each phase is often muddy, cyclical, and interconnected with one another, as UX designers very well know. Such was the case of the Defining and Ideating phases in this event. Both intertwined with each other by interlacing conversations between personal accounts of injustice and approaches to offset these often-painful experiences. In chapter 6, I examine three of the projects shared by the participants during the roundtable I moderated. I also draw on my notes to examine one of the projects I present in chapter 6.

Synthesizing

The traditional design thinking steps of Prototyping and Testing, as proposed by the Stanford d.school (2020), had already been done by the Indigenous interpreters and translators before attending the event. Instead of using the time together reinventing complex strategies and testing them days or weeks later,[2] participants drew on their own experiences and used these experiences as tested prototypes. We collectively revisited the issues and gathered takeaways from these experiences during the Synthesizing phase. We created a list of the most prominent issues faced by Indigenous interpreters and translators and the communities represented, reflecting on the strategies that have worked in the past to re-design ideas. Chapter 5 discusses this phase in detail.

Re-Designing

To some extent, the Re-Designing phase was our Prototyping phase. During this phase, we created a list of strategies that could be easily reproduced with the help of their own communities in order to bring potential solutions to the needs of Indigenous interpreters and translators and, thus, the needs of their communities. Some of these strategies were adaptations of what was presented, and some others were new ideas that sprouted from the discussions. Naming this phase Re-Designing rather than Prototyping seemed more suitable because, although we intended to create more tangible solutions, we realized that this is not always possible when working with complex issues such as the ones we did. Chapter 6 addresses the Re-Designing phase in detail.

Lastly, as stated before, even though this process seems linear, in practice it is a more cyclical process. We all come to a design thinking process with knowledges to contribute, and we then gain new knowledge that subsequently prompts us to do something with it. Thus, the last steps of our process became the beginning of new designs.

IRB APPROVAL

All participants gave their consent to CEPIADET to publish their experiences, which had already been published by CEPIADET on their YouTube channel (n.d.). Therefore, the request from the Institution Review Board (IRB) to conduct this research was submitted as a textual analysis of already-published raw data (FWA No. 00001224). In doing so, the raw data was preserved as Indigenous knowledge.

METHODS

To analyze and answer the research inquiries proposed in the introduction, this study uses qualitative methods, specifically semistructured interviews and testimonios. I also draw on my own observations and the notes I took throughout the event. The data gathered during the two days of the event is assembled in chapters 3 and 4 and synthesized in chapter 5. The semistructured interviews are coded in chapter 3, "Empathizing," and the testimonios in chapter 4, "Defining the Issues." I use pseudonyms to protect the privacy of the participants. All event affairs before, during, and after the unconference were conducted in Spanish, including the interviews and the testimonios, and thus are here translated.

Interviews

I analyze a set of twelve semistructured interviews, all with Indigenous interpreters and translators, conducted at the beginning of the event. All interviews were video recorded. I coded the data through user empathy maps (Wible 2020) that examine what participants say, do, think, and feel in order to find a deeper insight to their motivations, challenges, and needs (see table 2.1). Then, I identified the broader issues in separate categories to better understand the needs of Indigenous users. I provide more detailed information about this method in chapters 3 and 5.

Table 2.1. User empathy map adapted from Scott Wible (2020)

User:		Field:	
Place of Origin:		Languages:	
Motivations:	SAY (Quotes and Defining Words)	THINK (Thoughts and Beliefs)	Needs (Usability)
Challenges:	DO (Actions and Behaviors They Say They Do)	FEEL (Feelings and Emotions)	Needs (Deeper Meaning)

The interview protocol questions were as follows:

(Spanish)
1. ¿Cuál es su nombre? ¿De dónde viene? ¿Qué idiomas habla?
2. ¿Qué nos podría usted decir acerca de su profesión como intérprete y/o traductor/a?
3. ¿Qué le motivó a convertirse en intérprete y/o traductor/a de lenguas indígenas?
4. ¿Cuáles son los más grandes retos que enfrentan los/las intérpretes y/o traductores/as de lenguas indígenas diariamente?
5. ¿Qué le gustaría que la gente supiera acerca de las lenguas indígenas y de la interpretación de lenguas indígenas?

(English)
1. What is your name? Where do come from? What languages do speak?
2. Can you tell us a little about your background as an interpreter and translator?
3. Can you tell us about what motivated you to become an interpreter and translator of Indigenous languages?
4. What do you think are the biggest challenges that Indigenous language interpreters face in their daily work?
5. What would you like people to know about Indigenous languages and Indigenous language interpretation?

Through the user empathy map in table 2.1, I first identified the field, place of origin, and languages spoken by participants. Identifying this information was crucial to understanding the unique context of each participant. On the left side of the table, I placed a space to add the participants' motivations and challenges, which the participants specified in their interviews as they were directly asked to identify these two aspects of their profession. In the middle of the table, I placed a space to identify what participants said during the interviews, what they thought (based on the interview quotes), what they said they do in their profession, and how they felt (also based on what the participants said during their interview). The right side of the table was used to identify the needs of each participant; first, their usability needs (more material needs) at the top right side of the table, and then their deeper meaning needs at the bottom right side. Analyzing interviews through user empathy maps allowed me to deconstruct each interview for an in-depth analysis of each experience.

Testimonios

I analyze nine testimonios, all of which were shared during the roundtable discussion I moderated (although twelve people participated in the roundtable I moderated, only nine shared their testimonios). The testimonios include seven testimonios from Indigenous interpreters and translators, one testimonio from a non-Indigenous social anthropologist and human rights professor from Peru, and one testimonio from a translation and interpreting studies professor also from Peru who works with Indigenous translators in that country but teaches at a university in Belgium. Instead of the researcher asking one question at a time, as in the case of interviews, testimonios draw on prompts to build a narrative and produce dialogical conversations that seek to understand the perspective of a group of people. The two main prompts in this study included the following:

1. ¿Nos puede decir su nombre, las lenguas que habla y de dónde nos visita? (Can you tell us your name and your background?)
2. ¿Qué nos puede decir acerca de los desafíos que tiene usted como intérprete y traductor de lenguas Indígenas? (What can you tell us about the issues you face as an interpreter and translator of Indigenous languages?)

The first prompt initiated a protocol of introduction. Each participant took about five minutes to introduce and provide contextual background to the rest of the group. During the second prompt, participants discussed issues affecting their profession in their communities and contributed with ideas they have tested to confront the problems. Each participant took about fifteen minutes. At the end of each contribution, the group engaged in dialogical, reflective conversations about what was presented. All testimonios were audio recorded. I coded the information in testimonios by mapping the individual and collective *pain points*—specific problems experienced by the users (Stanford d.school 2020)—in each testimonio with the purpose of defining the

Table 2.2. Testimonio map

Name:	Indigenous Translator/Interpreter:
Place of Origin:	Languages:
Field:	

Pain Points:	Issues Identified:

Civic Engagement:	Ideas Tested:

Outcomes:	Implications:

problems faced by Indigenous interpreters and translators before, during, and after T&I events (see table 2.2). It should be noted that three of the twelve interviewees are also part of the nine participants sharing testimonios. As in the interviews, I identify the broader issues in separate categories for a comprehensive examination of the needs of Indigenous users. I provide more detailed information about this method in chapters 4 and 5.

Like in the user empathy maps, testimonio maps also traced the place of origin, the field, and the languages spoken by the participants in order to understand each unique context. In these maps, however, I thought it pertinent to add a field to identify whether the participant self-identified as Indigenous because not all participants who shared their testimonios did. This table was divided into three rows that directly stem from the arc of testimonios, which are narratives that discuss a personal experience that exposes a collective struggle that in turn promotes civic engagement (Rivera 2022b). For example, in the first row I identified individual and collective pain points on the left, which, as explained before, are specific issues faced by users. I used direct quotes to identify these pain points, and then extracted thematic issues from them, which I placed on the right side of the first row. Then in the second row on the left I identified the civic engagements in which the participants take part based on what they quoted during their testimonios. From these engagements, I extracted thematic strategies and ideas that participants have previously tested. I placed these tested ideas on the right side of the second row. I used the last row to identify the outcomes of their civic engagements on the left and the implications of these outcomes on the right. While the outcomes and implications were not always clearly stated by the participants, each civic engagement yielded important implications for Indigenous professionals in the field of interpreting and translation.

AUDIO-TO-TEXT TRANSCRIPTIONS AND TRANSLATIONS

Because all interviews and testimonios were recorded in Spanish, I transcribed and translated each one of them. I decided not to use an online service because the sound of some of my recordings was not of high quality, and some of the audio was difficult to hear. Also, because of the different Spanishes spoken by the participants and the different areas and fields in which they work, I knew that I was going to have to conduct additional research about specific terms, concepts, town names, institutions, and organizations. Although as a Spanish-English bilingual educator I often help in matters of interpreting and/or translating, I am not a professionally trained translator or interpreter. Therefore, my translations of the interviews and testimonios might not have the quality standards they should have. To ensure that my translations were consistent, and to make my process a little easier to manage, I created a short glossary of technical terms that reappeared often in the various interviews and testimonios (see table 2.3). Also, a family member who has practiced law in both Mexico and the United States helped me to identify the correct translations of legal concepts and terms.

In my glossary, I identified terms that repeatedly appeared in both the interviews and the testimonios. I did this to ensure that my translation was consistent throughout this study. On the left of my glossary, I placed the terms in Spanish. Then, I placed the concepts in English on the right side of the table. This exercise gave me a glimpse of the complex job of professional translators, which I have always admired and respected. Creating technical glossaries in translation exercises is challenging because specialized terminology is highly localized in the context of the geographic spaces where the terms are used.

The International Unconference for Indigenous Interpreters and Translators brought together a group of participants who work in different countries, in various professional contexts, and with different languages, embodying the interdisciplinary, multicultural, and multilingual nature of the field of translation and interpretation studies. For this reason, analyzing the event through a localized design thinking process lens is important. This adapted approach emphasizes an Indigenous lens that analyzes a design thinking process that (1) empathizes through interviews and exploration of the backgrounds of Indigenous interpreters and translators; (2) defines the issues through testimonios that emphasize the use of dialogue and desahogo as Indigenous practices; (3) ideates through ideas that Indigenous interpreters and translators

Table 2.3. Spanish-English glossary

Spanish Term	English Term
mantener	to sustain
resiliencia	resilience
sensibilizar	to create awareness
asociación civil	nonprofit/nonprofit organization
consulta previa	prior consultation/consultation
usos y costumbres	customs and traditions
viáticos	travel expenses
juicios orales	oral proceedings
audiencia	hearing
proceso anticipado	anticipated process/plea deal/plea bargain
conmutación de la pena	commutation of sentence
actuario	(court) process server
perito	surveyor

have tested in the past; (4) synthesizes by reflecting on the needs and the issues with which Indigenous interpreters and translators grapple; and (5) re-designs, rather than prototypes, strategies that can be used to help professionalize the field of Indigenous interpreting and translation. The following chapter details the Empathizing phase of the design thinking process traced during the event.

3
EMPATHIZING

A CRITICAL VIEW OF EMPATHY

When I was invited to collaborate in the coordination of the first International Unconference for Indigenous Interpreters and Translators by my good friend and colleague Dr. Monica Morales Good, I hesitated about my participation as a collaborator. Having read a fair share of articles and essays by Indigenous scholars who understandably see with skepticism the work of non-Indigenous scholars discussing Indigenous matters, I questioned my own positionality as a Mestiza and doubted my right to work with Indigenous communities. As an educator, I have worked with students and colleagues from different parts of the world, and from various linguistic backgrounds. Yet, none of these interactions have come to me as politically charged as working side by side with Indigenous people. I felt—feel—in part accountable for the systemic issues that they grapple with, which are caused by the appropriation of their lands, the racist attitudes lurking in our societies, and the westernized systems led by Mestizes in Latin America or by Anglos in the United States and Canada. I accepted my friend's invitation because I felt that if Mestizes had been part of the problem, then we could be part of the solution. Even though I neither live in Indigenous communities nor speak their languages, I am convinced of the many benefits of building alliances between cultures that sprouted from the same tree, like learning from one another and helping one another grapple with the many systematic issues of our times *together*.

Empathy is commonly understood as the attempt to understand someone else's point of view by placing oneself in the situation of others. But in truth, one can never really walk in the shoes of others because each human experience is unique. I can never entirely comprehend the experiences of Indigenous interpreters and translators. Yet, I can confront my own prejudices and uncover the many similarities between us. This work proposes to expand our understanding of empathy through a critical lens by reflecting on and questioning a researcher's position in a project. *Critically empathizing* builds bridges of understanding by

https://doi.org/10.7330/9781646425310.c003

confronting our own prejudices, acknowledging our differences, and uncovering similarities between the researcher(s) and participants. Critical empathy is complex and uncomfortable as it carries out well beyond the duration of a research project. In my case, I have to repeatedly explain my identity and why I work with Indigenous communities to audiences unfamiliar with Latin American contexts, and, most important, I often reflect on these collaborations and inquire about what else I could do to continue building bridges of understanding between Indigenous and Mestize groups. Although empathizing with Indigenous interpreters and translators is not enough, it is an important step to begin to comprehend systematic problems from an Indigenous point of view. Empathy, as explained by the Stanford d.school (2020),

> is the centerpiece of a human-centered design process. The Empathize mode is the work you do to understand people, within the context of your design challenge. It is your effort to understand the way they do things and why, their physical and emotional needs, how they think about the world, and what is meaningful to them.

To empathize in a UX research project, we observe, engage, and listen through interviews or shadowing in order to seek nonjudgmental understanding (Stanford d.School 2020). Empathy, nonetheless, is often romanticized as an idealistic and static, hollow concept devoid of action (Tham 2022). It is easy to overlook that *empathizing* is an active verb that extends an invitation to all individuals participating in the research, including the researcher(s), to act upon "ways to mitigate racial, ethnic, and social discrimination" (274). Empathizing problematizes the so-called researcher neutrality, forcing us to reflect on our biases and prejudices as individuals brought up in Eurocentric institutions.

While empathizing is identified as the first step of the design thinking process, in the case of our event, the act of empathizing took place before, during, and after the event. At every point, all of us who participated in the event were compelled to (1) question our own prejudices, such as when I noticed that the verbal abilities of most Indigenous participants exceed the verbal abilities of many scholars—including myself; (2) recognize our differences, like when one of the Indigenous participants referred to the Mexican court as the "Mestizo court," a term that forced me to see that my Mestiza lens was clearly different from the Indigenous lens; and (3) uncover similarities among the numerous cultures and languages represented in the event, such as when I learned that many Indigenous interpreters began interpreting for their family members as children, similarly to the experiences of many of the Mestize students I have taught in US schools.

INDIGENOUS INTERPRETERS AND CHILD LANGUAGE BROKERING

Like many interpreters of minoritized languages, most Indigenous interpreters become involved in this profession as a result of the need to interpret as children for a family member who did not speak the language spoken by public officials and/or the staff members of public institutions, as many of the participants in the event pointed out. Highlighting this background helps us see a more complete picture of the motivations and the needs of Indigenous interpreters and translators. Victoria, for example, explained that she became an interpreter and translator to help her monolingual mother navigate the legal proceedings after her father died in a traffic accident (see table 3.4). Similarly, Lucas became an interpreter to help his mom during her doctor's appointments when he was in middle school, and "because [he] translated for [his] mom, the doctor called [him] to translate for other patients who didn't speak Spanish" (see table 3.11).

Chicana feminist historian Antonia Castañeda (1998) argues that "children play an adult role while they are translating" and believes that we ought to examine the experience of child translators closely (299). Castañeda contends that "childhood's boundaries are transgressed each time a child is required to translate—and thus mediate, negotiate, and broker adult realities across cultures" (231). In her seminal work "Language and Other Lethal Weapons: Cultural Politics and the Rites of Children as Translators of Culture," Castañeda narrates her experience as a child language broker at the age of seven:

"Dile que no puedo respirar—que se me atora el aire. Dile . . ."
How do I say "atora"?

"Tell your mother that she has to stop and place this hose in her mouth and press this pump or else she will suffocate."

"¿Qué dice? ¿Qué dice?"
He was sitting behind this big desk, and my mother was sitting beside me and holding onto my hand very tightly.

I . . . what does suffocate mean? How do I translate this? I don't have the words.

"¿Qué dice? ¿Qué dice?"

"I . . . uh . . . Dice que . . . wh . . . Dice que si no haces lo que te dice te mueres." (229)

Castañeda's account exemplifies the terrifying situation of holding her mother's life in her tongue as a child of only seven years old, a common situation for bilingual children of migrant parents and bilingual children of monolingual Indigenous parents. As Claudia Angelelli

(2016) asserts, the practice of allowing ad-hoc "family interpreters" problematizes the policies of the public industry. Although interpreting at a young age may enhance the confidence of bilingual children, this practice places massive stress on children who have to handle issues for which they are not sufficiently matured (Angelelli 2016). Another illustration of this is the experience narrated by Amanda during her interview (see table 3.12):

> My grandfather died, and my grandmother was involved in a legal process.... My grandmother only spoke Quechua.... Every time she went to court, no one understood her.... I was eight years old when my grandmother told me, "Let's go, you'll accompany me." I had to interpret from my grandmother to the judge and also from the public defendant to my grandmother. Amid this dilemma, I understood that there was an oath involved too, which was intimidating. They said, "you have to tell the truth of what she is saying, and only that."

At only eight years old, Amanda not only had to learn to interpret in the context of the painful situation in which she and her grandmother were part after the loss of her grandfather but she also had to go through the traumatic experience of navigating a public sector's cosmovision that threatened her if she did not swear an oath that she did not understand. In addition to the ethical implications that these practices hold, there could also be serious legal and medical consequences for the party for whom the child interprets if it is not done correctly. This contextual upbringing of Indigenous interpreters is crucial to understand their motivations and their needs.

MAPPING INTERVIEWS

During the International Unconference for Indigenous Interpreters and Translators, I conducted twelve interviews with Indigenous interpreters and translators with the help of my colleagues. Because all participants were involved in the different roundtable conversations that were taking place during the event, volunteers were asked to step outside the roundtables for eight to ten minutes to answer a series of five questions related to their profession. As discussed in chapter 2, the interviews gathered information about their motivations and their challenges in order to map what they say, think, do, and feel in user empathy maps (Wible 2020). The maps allow us to clearly identify the needs of Indigenous interpreters and translators both at the surface level and at a deeper level. On the one hand, as Scott Wible (2020) denotes, the needs at the surface level are material needs that mainly address issues of usability. In the case of Indigenous interpreters, these material needs stem directly from their profession, like professionalization, follow-up training, and dignified wages. On the other hand, there are also deeper-level needs revealed in these maps, which "foreground insights about the emotional depth and breadth of a person's experiences rather than only their material needs" (Wible 2020, 413). Deeper-level needs revealed by Indigenous interpreters and translators are almost always related to their Indigenous identities and their sense of responsibility to their communities, like helping their communities by promoting awareness about Indigenous cultures and about Indigenous language rights. Although I separate usability needs from deeper-level needs in these maps in order to deconstruct the experiences of Indigenous interpreters and translators, both usability needs and deeper-level needs are strongly interconnected because, as I discuss in chapter 5, most Indigenous interpreters and translators see their profession as part of their responsibility to their community.

The interviews were video recorded, and each interview took approximately five to eight minutes. All the interviews were conducted in Spanish. Instead of using a digital service, I transcribed and translated the information in the interviews manually, mainly because I knew that I had to research some of the information given to ensure that names and concepts were appropriately transcribed and translated within their contexts. The names of the communities to which the interviewees belong and the names of institutions and/or processes that some interviewees mentioned could have easily caused problems in a digital transcription and/or translation. It is also important to note that these interviews were unrehearsed, and unlike transcribing and translating rehearsed material, such as presentations and/or podcast interviews, these interviews had the dialogical characteristic of unrehearsed oral communication. Therefore, it would have been difficult for a digital transcription and/or translation software to identify minutiae important to this study, such as who "them" and "us" referred to in the interviewees' responses.

Moreover, I used sticky notes, a strategy often used by design thinkers (Wible 2020), to easily shift individual responses into collective responses in order to identify the motivations, challenges, usability needs, and deeper-level needs of the group (see figures 3.1 and 3.2). I found this seemingly simplistic strategy highly malleable and helpful in the process of identifying collective needs, which are detailed in chapter 5. In a traditional UX research project, this strategy with sticky notes is done as a group activity with the participants. However, as stated before, the roundtables and the event overall were not originally designed as a UX project,

46   EMPATHIZING

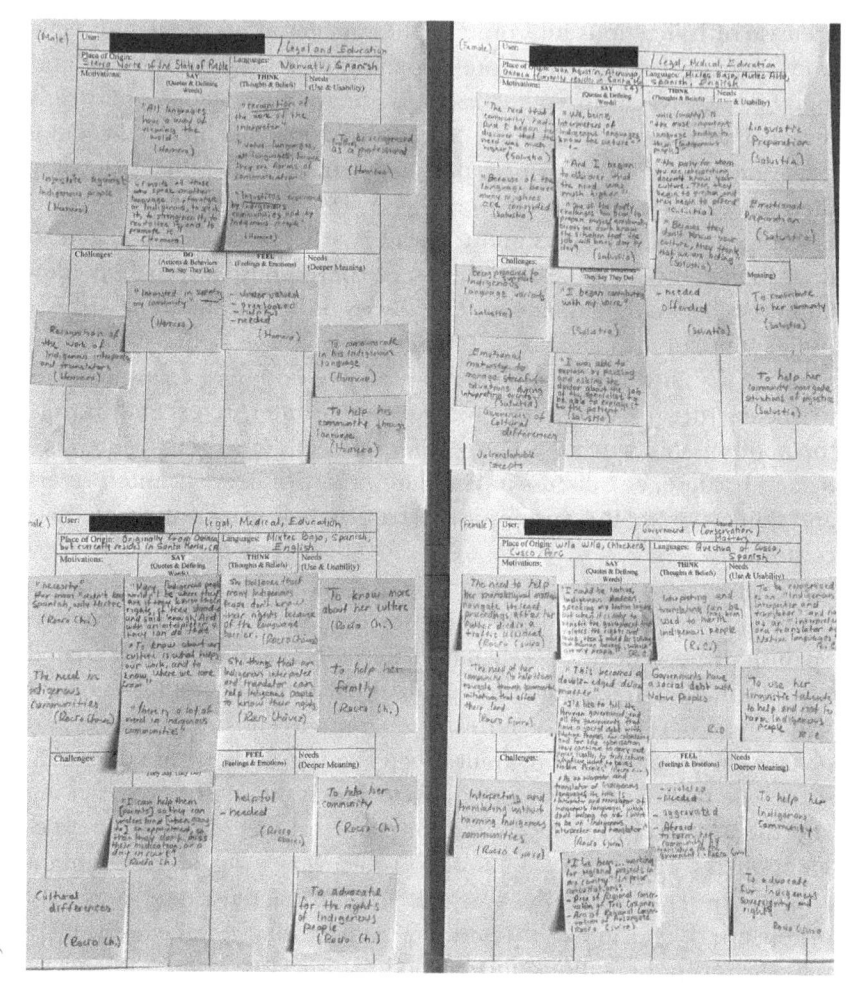

Figure 3.1. Examples of individual user empathy maps

as we followed the lead of the Indigenous participants and not the other way around. Also, some participants might have felt uncomfortable using the Spanish language in its written form, an issue not always considered by UX researchers when working with multilingual participants.

Figure 3.1 shows four examples of how sticky notes were used to code each user's empathy map. Once all analog empathy maps were coded, I transcribed each into individual tables, as shown in the next section. Figure 3.2 shows how I collected in categories all the information given by the Indigenous interpreters and translators who participated in the interviews to form a *collective* empathy map. Each category mimicked the

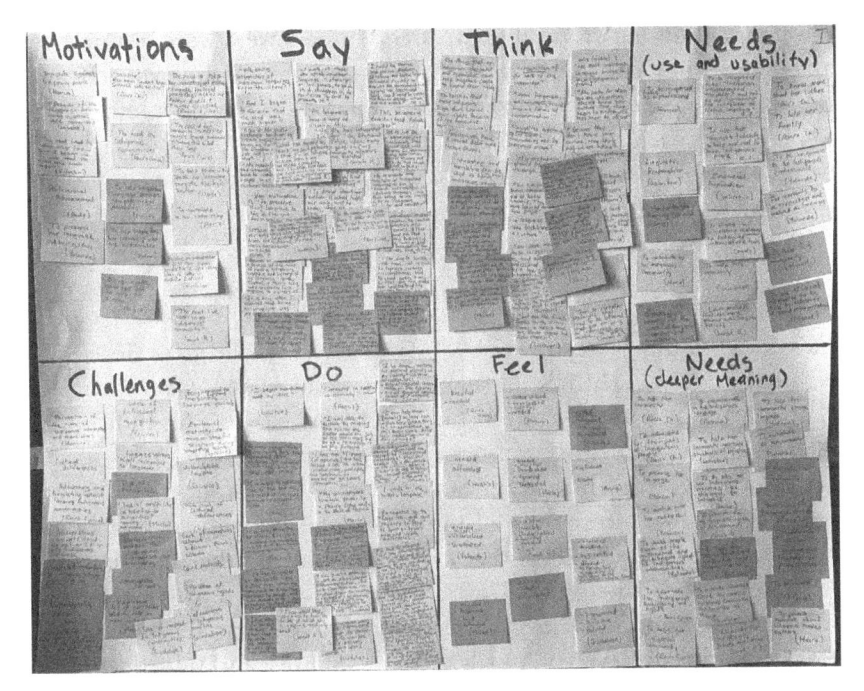

Figure 3.2. Collective empathy map

sections in the user empathy map: Motivations, Challenges, Say, Think, Do, Feel, Usability Needs, and Deeper-Meaning Needs. Because I had placed the information on sticky notes (with the name of each participant on each sticky note), this task was easy to do.

It should be noted that the color of the sticky notes bears no relation to how I conducted my research. I used the colors of sticky notes available to me. Once all the information was grouped by category, it was easy to identify themes within each category. And because each sticky note had the name of the participant who provided that specific information, it became easy to identify who said what during the synthesizing process.

USER EMPATHY MAPS

Each empathy map was transcribed into individual tables to easily identify a pattern. The contents of the Motivations, Challenges, Say, Think, and Do sections in the maps consist of a combination of direct quotations from the interviewees and my own statements summarizing the information the interviewees gave. The contents of the Feel and Needs sections are my own interpretations of the feelings and the needs of the interviewees based on their answers. Interviewees were not directly asked how they felt about their work as Indigenous interpreters and translators or what they thought their immediate and deeper-level needs were. Before each user empathy map, I provide a summary of each participant's experience based on the map. Pseudonyms are used to protect the privacy of the participants.

Luis

Luis lives in the State of Puebla, Mexico (see table 3.1). He speaks Nahuatl and Spanish. His journey began as an elementary teacher of Nahuatl, and he later became a court interpreter to help his Indigenous community navigate the injustices they experience. He works in the courts of the Sierra Norte in the State of Puebla. His main challenge has been the lack of recognition of his work as an Indigenous interpreter and translator. Luis believes that it is important to value all languages. Although he feels helpful and needed by his community, he also feels overlooked and undervalued as an Indigenous interpreter and translator. At the surface level, he needs to be recognized and valued as a professional. At a deeper level, he needs to communicate in his Indigenous language. Luis also feels the need to help his community through language.

Table 3.1. Luis's empathy map

User: Luis		Field: Legal, Educational	
Place of Origin: Sierra Norte of the State of Puebla (Mexico)		Languages: Nahuatl, Spanish	
Motivations:	SAY (Quotes and Defining Words)	THINK (Thoughts and Beliefs)	Needs (Usability)
Injustice against Indigenous people	"All languages have a way of viewing the world." "I invite all those who speak another language, foreign or Indigenous, to speak it, to strengthen it, to revitalize it, and to promote it."	His daily challenge is "the recognition of the work of the interpreter." He believes that it is important "to value all languages, because they are forms of communication." He believes that there are "injustices experienced by Indigenous people."	To be recognized and valued as a professional

continued on next page

Table 3.1. Luis's empathy map—*continued*

CHALLENGES:	DO (ACTIONS AND BEHAVIORS THEY SAY THEY DO)	FEEL (FEELINGS AND EMOTIONS)	NEEDS (DEEPER MEANING)
Recognition of the work of Indigenous interpreters and translators	"At first, I studied to become a teacher of Indigenous education at the elementary level in the area of Nahuatl, but then I was interested in supporting my community." "I have supported the courts of the Sierra Norte of Puebla with interpreting and translation in Indigenous languages, mainly Nahuatl and Spanish."	Undervalued Overlooked Helpful Needed	To communicate in his Indigenous language To help his community through language

Mariana

Mariana is originally from San Agustín Atenango, Oaxaca, Mexico, but she currently resides in Santa María, California (see table 3.2). She speaks Mixteco Bajo (from the lowlands), Mixteco Alto (from the highlands), Spanish, and English. She works in the legal, educational, and medical fields in California. Like most Indigenous interpreters, she became involved in this profession after seeing the many injustices committed against her community members because of the language barrier. Her most significant challenges have been the lack of awareness of Indigenous language variants, lack of awareness of cultural differences, managing untranslatable words during interpreting events, and managing emotions while interpreting stressful situations. She believes that contributing with her voice is important because orality is the most important communication tool for Indigenous people. She also believes that public officials often end up offending Indigenous people because of a lack of awareness of cultural differences. Mariana feels needed by her community, but she also has felt offended by public officials who do not understand her culture. Her usability needs include acquiring linguistic preparation regarding the variants she speaks and acquiring emotional preparation to handle stressful situations. Her needs at a deeper level include promoting cultural awareness and contributing to her community by helping her people navigate situations of injustice.

Table 3.2. Mariana's empathy map

USER: Mariana	FIELD: Legal, Educational, Medical
PLACE OF ORIGIN: Originally from San Agustín Atenango, Oaxaca (Mexico) but currently resides in Santa María, California	LANGUAGES: Mixteco Bajo, Mixteco alto, Spanish, English

MOTIVATIONS:	SAY (QUOTES AND DEFINING WORDS)	THINK (THOUGHTS AND BELIEFS)	NEEDS (USABILITY)
Seeing the need in the community "Because of the language barrier, many injustices are committed."	"We, as interpreters of Indigenous languages, know the culture." "And I began to discover that the need was much higher." "One of the daily challenges has been to prepare emotionally because we don't know the situation that the job will bring day by day."	"I started to contribute with my voice, the most important communication bridge to them [Indigenous people]." "The party for whom you are interpreting doesn't know your culture. They begin to question, and they begin to offend. Because they don't know our culture, they think that we are acting, but in reality, they don't know our culture."	To acquire linguistic preparation regarding variants To acquire emotional preparation To promote cultural awareness among public officials

CHALLENGES:	DO (ACTIONS AND BEHAVIORS THEY SAY THEY DO)	FEEL (FEELINGS AND EMOTIONS)	NEEDS (DEEPER MEANING)
Being prepared to support Indigenous language variants Emotional maturity to manage stressful situations during interpreting events Lack of awareness of cultural differences Untranslatable concepts	"I started to contribute with my voice." "I was able to explain by pausing and asking the doctor about the job of the specialist to be able to explain it to the patient."	Needed Offended	To contribute to her community To help her community navigate situations of injustice

Gabriela

Gabriela is originally from Oaxaca, Mexico, but currently resides in Santa María, California (see table 3.3). She speaks Mixteco Bajo (from the lowlands), Spanish, and English. She became an interpreter to help her parents with doctor appointments and immigration court hearings. She works in the legal, educational, and medical fields. She believes that many Indigenous immigrants don't know their rights because of the language barrier and that having an interpreter can help them. She feels helpful and needed by her family and her community. Her usability need is to know more about her culture to be able to perform better interpretations. Her deeper-level needs include helping her family, helping her community, and advocating for Indigenous rights.

Table 3.3. Gabriela's empathy map

User: Gabriela	Field: Legal, Educational, Medical
Place of Origin: Originally from Oaxaca (Mexico) but currently resides in Santa María, California	Languages: Mixteco Bajo, Spanish, English

Motivations:	SAY (Quotes and Defining Words)	THINK (Thoughts and Beliefs)	Needs (Usability)
The need for her mom to communicate The need she sees in Indigenous communities	"Many [Indigenous people] wouldn't be where they are if they knew their rights, if they stood up and said 'enough.' And they can do that with an interpreter." "To know about our culture is what helps our work, and to know where we come from." "There is a lot of need in Indigenous communities."	She believes that many Indigenous people don't know their rights because of the language barrier. She thinks that having an interpreter can help Indigenous people know their rights.	To know more about her culture

Challenges:	DO (Actions and Behaviors They Say They Do)	FEEL (Feelings and Emotions)	Needs (Deeper Meaning)
Cultural differences	"I became an interpreter out of necessity. Because of what I went through with my parents, with my mom more than anything, because she didn't know Spanish, only Mixtec, and every time we went to an appointment, she didn't know how to [express herself]. She only used signs to point out the schedule to go [to appointments]. And now that I know Mixtec, Spanish, and English, I can help them so that they can understand [when going to] an appointment, so that they don't miss their medication, or a day in court."	Helpful Needed	To help her family To help her community To advocate for Indigenous rights

Victoria

Victoria is from Wila Wila, Chinchero, Cusco, Peru (see table 3.4). She speaks Quechua (from Cusco) and Spanish. She became an interpreter to help her monolingual mother navigate the legal proceedings after her father died in a traffic accident. She has worked on several "prior consultations," which are part of a conservation policy of the Peruvian government that began in 2013 that requires government agencies and private organizations to "consult" with Indigenous Peoples before taking on developmental initiatives close to their communities. She believes that interpreting and translating can be, and has been, used to harm Indigenous people. In her interview, she cites "Felipillo," an Indigenous interpreter who helped the Spaniards conquer her region.[1] She thinks that governments have a social debt to Native Peoples for the colonization of their lands, colonization that continues today. She feels that her rights have been violated. And although she feels needed as an interpreter, she is also afraid to harm her community by translating and interpreting for the government. Her usability need is to be recognized as an "Indigenous interpreter and translator" and not as an "interpreter and translator of Native languages." Her deeper-level needs include using her linguistic talents to help and not harm her Indigenous community and advocating for Indigenous rights and sovereignty.

Table 3.4. Victoria's empathy map

User: Victoria		Field: Government (land conservation matters)	
Place of Origin: Wila Wila, Chinchero, Cusco, Peru		Languages: Quechua of Cusco, Spanish	
Motivations:	SAY (Quotes and Defining Words)	THINK (Thoughts and Beliefs)	Needs (Usability)
The need to help her monolingual mother navigate the legal proceedings after her father died in a traffic accident. The need to help her community navigate governmental initiatives that affect their land	"I could be Native Indigenous Andean, speaking my Native language, but when it is only to benefit a government that violates the rights and laws, then I would be selling out human beings, which are my people." "This [job] is a double-edged delicate matter."	She believes that interpreting and translating can be, and has been, used to harm Indigenous people. In the interview she cited "Felipillo," an interpreter who helped the Spaniards conquer her region, similar to the figure of La Malinche.	To be recognized as an "Indigenous interpreter and translator" and not as an "interpreter and translator of Native languages"

continued on next page

Empathizing 53

Table 3.4. Victoria's empathy map—*continued*

MOTIVATIONS:	SAY (QUOTES AND DEFINING WORDS)	THINK (THOUGHTS AND BELIEFS)	NEEDS (USABILITY)
	"I'd like to tell the Peruvian government and all the governments that have a social debt to Native Peoples for colonizing and for continuing to carry out colonization now, legally, to truly return what we want as Native Peoples." "The title, 'Interpreter and Translator of Native Languages,' doesn't belong to me. I want to be called, 'Indigenous Interpreter and Translator.'"	She thinks that governments have a social debt with Native Peoples.	

CHALLENGES:	DO (ACTIONS AND BEHAVIORS THEY SAY THEY DO)	FEEL (FEELINGS AND EMOTIONS)	NEEDS (DEEPER MEANING)
Interpreting and translating without harming Indigenous communities	"I've been . . . working for regional projects in my community." She said she has worked in "prior consultations" in the Area of Regional Conservation of Tres Cañones and in the Area of Regional Conservation of Ausangate.	Violated Needed Aggravated Afraid to harm her community by translating and interpreting for the government	To use her linguistic talents to help and not to harm her community To advocate for Indigenous rights and sovereignty

Pedro

Pedro is from Miahuatlán de Porfirio Díaz, Oaxaca, Mexico (see table 3.5). He speaks Zapoteco del Sur (from the Southern part of the state) and Spanish. Pedro began working for the court system as a student intern. When his employer found out that he spoke Zapotec, he was asked to interpret for Zapotec defendants and victims. He now works as a court interpreter. He believes that interpretation should be made through dialogue because many legal concepts do not exist in his Indigenous language. He is motivated by professional advancement and feels confident about his abilities. He also feels needed by his employer and by Zapotec speakers. His usability needs are to engage in interpreting dialogues to construe untranslatable terms. At the deeper level, Pedro needs to be able to use Indigenous practices, like dialogue, to interpret information effectively.

Table 3.5. Pedro's empathy map

User: Pedro	Field: Legal
Place of Origin: Miahuatlán de Porfirio Díaz, Oaxaca	Languages: Zapoteco del Sur, Spanish

Motivations:	SAY (Quotes and Defining Words)	THINK (Thoughts and Beliefs)	Needs (Usability)
Professional advancement	"I introduced myself to this court . . . to conduct my internships and social service. After they found out that I spoke this Indigenous language, I was asked to help translate for defendants and victims." "You try to speak with them, not so much to explain certain circumstances but to chat with them, so that you can make them understand the circumstances."	He believes that interpretation should be done through dialogue.	To engage in interpreting dialogues to interpret untranslatable terms

Challenges:	DO (Actions and Behaviors They Say They Do)	FEEL (Feelings and Emotions)	Needs (Deeper Meaning)
Interpreting concepts that do not exist in his Indigenous language	"An Indigenous person cannot comprehend the concept of 'summary proceeding,' then you explain, 'it is the process or the way,' but because I won't be able to explain the concept of 'process' either, 'it is the way in which they will help you if you accept that you committed a crime,' then again I won't be able to say 'crime,' 'that you committed an action against justice, a bad action,' you tell him, 'and if you accept your responsibility, the judge will give you the minimum sentence.' You won't be able to explain 'minimum' either, so you say, 'the least they can.'"	Confident Needed	To be able to use Indigenous practices, like dialogue, to interpret

Natalia

Natalia is from Peru, from the community of Aiza, district of Tupe, province of Yauyos, city of Lima (see table 3.6). She speaks Jaqaru, an Andean language, and Spanish. She is an Indigenous linguist, translator, and interpreter. One of the highlights of her career is being an Indigenous translator in the first bilingual (Spanish-Jaqaru) civil registry program sponsored by the Peruvian government in 2014. She translated birth and death certificates to accommodate the linguistic needs of the Jaqaru communities. Her biggest challenges have been the lack of awareness about Indigenous linguistic variants as well as the lack of professional recognition as an Indigenous linguist, translator, and interpreter. Although she feels needed by her Indigenous community, she also feels undervalued and overlooked by the public sector. At the surface level, she wants to be valued as an Indigenous professional and create awareness about Indigenous language variants. Her deeper-level needs include preserving her language and culture and promoting awareness about the multicultural and multilingual aspects of Indigenous communities.

Table 3.6. Natalia's empathy map

User: Natalia		Field: Education, Government (Civil Registry)	
Place of Origin: Aiza from the district of Tupe, province of Yauyos, Lima, Peru		Languages: Jaqaru (Andean language), Spanish	
Motivations:	SAY (Quotes and Defining Words)	THINK (Thoughts and Beliefs)	Needs (Usability)
To preserve her language and her culture	"My reason [to translate and interpret] has always been to watch over my people in all its cultural aspects." Her motivation is "to preserve [her] language in its different forms." "There are many variants that differ from one another, and these variants should be respected when people translate."	"Not all know alphabetic literacy in our Indigenous or Native communities. This is still limited to very few people." She believes that "a language is the backbone of a culture." She thinks that her work "needs to be recognized as any other professional working as a translator or interpreter."	To be valued as an Indigenous professional To create awareness about Indigenous language variants
Challenges:	DO (Actions and Behaviors They Say They Do)	FEEL (Feelings and Emotions)	Needs (Deeper Meaning)
Language variants aren't recognized as languages Lack of professional recognition	"I have training in linguistics, and as a linguist, I've always been conscious about the value of my language, which is the backbone of a culture." "I write in the Jaqaru language." She participated in the "first bilingual civil registry in Peru" of Spanish-Jaqaru birth certificates and death certificates.	Needed by Indigenous people Helpful to Indigenous people Undervalued by the public sector Overlooked by the public sector	To preserve her language To watch over her culture To promote awareness about the multicultural and multilingual aspects of Indigenous communities

Claudia

Claudia is originally from the State of Chiapas, Mexico, but currently resides in Mexico City (see table 3.7). She speaks Tzeltal de los Altos de Chiapas (the highlands of Chiapas) and Spanish. She became a court interpreter to help her Tzeltal community in Mexico City, which is seen as an Indigenous immigrant group in that city. She has worked as an interpreter and translator for over fifteen years. She sees her profession as a calling because she uses her language to help her community. She feels needed by her community but ignored and neglected by the public systems in Mexico. Her usability need is to create awareness among public authorities about Indigenous cultures. Her deeper-level needs include creating awareness about Indigenous Peoples' matters and contributing to her community.

Table 3.7. Claudia's empathy map

User: Claudia		Field: Legal	
Place of Origin: Originally from the State of Chiapas, Mexico, but currently resides in Mexico City		Languages: Tzeltal de los Altos de Chiapas, Spanish	
Motivations:	SAY (Quotes and Defining Words)	THINK (Thoughts and Beliefs)	Needs (Usability)
To help those who speak her language navigate the legal system To contribute to her community	She became an interpreter because she has "Indigenous brothers and sisters who speak Tzeltal in Mexico City constantly facing problems in matters of justice." "It is here where I realized that being an interpreter was my calling."	She sees interpreting as a calling. She believes that "there's still much ignorance about Indigenous Peoples' issues." She thinks that "there is a lack of awareness, perhaps even ignorance, about the current laws that benefit Indigenous Peoples."	To create awareness among public authorities about Indigenous cultures
Challenges:	DO (Actions and Behaviors They Say They Do)	FEEL (Feelings and Emotions)	Needs (Deeper Meaning)
Lack of sensibility on behalf of the authorities regarding Indigenous cultures Lack of awareness about Indigenous Peoples issues	"I have over 15 years [of experience] and have learned a lot. We continue collaborating, participating, and contributing to benefit our Indigenous communities." "My interventions have been primarily in Mexico City and in the State of Mexico."	Needed Helpful Overlooked Ignored Neglected	To create awareness about Indigenous Peoples matters To contribute to her Indigenous community

Alejandro

Alejandro is from Miramar, municipality of Santa María Yucuhiti, district of Tlaxiaco, Oaxaca, Mexico (see table 3.8). He speaks the Mixteco del Noroeste variant and Spanish. He works as a court interpreter. He is motivated by helping Indigenous language speakers navigate the legal

Table 3.8. Alejandro's empathy map

User: Alejandro	Field: Legal
Place of Origin: Miramar, municipality of Santa María Yucuhiti, district of Tlaxiaco, Oaxaca (Mexico)	Languages: Mixteco del Noroeste variant, Spanish

Motivations:	Say (Quotes and Defining Words)	Think (Thoughts and Beliefs)	Needs (Usability)
To help Indigenous language speakers navigate legal processes	"I got involved in this job because of the many people who speak an Indigenous language who are deprived of their freedom and don't have anyone to help them interpret or translate documents or hearings in the legal processes they are in."	"I act as a communication bridge between the Mixtec language and Spanish and vice versa, so that the process is understood by the Indigenous person." "There are still statewide and other authorities that do not provide dignified wages for Indigenous interpreters."	Dignified wages for the work of Indigenous interpreters and translators

Challenges:	Do (Actions and Behaviors They Say They Do)	Feel (Feelings and Emotions)	Needs (Deeper Meaning)
Unfair wages for Indigenous interpreters and translators. Lack of recognition for Indigenous people who work as interpreters and/or translators	"I am the Coordinator of Interpreters and Translators of Indigenous Languages [at CEPIADET]. I take care of matters regarding the requirements by the courts and the management of the interpreters' payments." Besides interpreting in the courts, he has translated for government institutions, such as Instituto Nacional de Lenguas Indígenas (INALI), Inter-American Institute for Global Change Research (IAI), and Instituto Nacional para la Educación de los Adultos (INEA).	Needed Helpful Indignant Offended Disrespected	Validation for Indigenous interpreters and translators as professionals To be valued

systems in Oaxaca. He sees his profession as a "communication bridge." As the coordinator of Indigenous interpreters and translators at CEPIADET, he sees unfair wages and lack of professional recognition as the most significant challenges faced by Indigenous court interpreters. Although Alejandro feels helpful and needed by his community, he also feels offended and disrespected by the public system. His surface-level need is dignified wages for the work of Indigenous interpreters and translators, and his deeper-level need is validation for Indigenous professionals.

Samuel

Samuel is from Mazatlán Villa de Flores, Oaxaca, Mexico (see table 3.9). He speaks Mazateco and Spanish. He works as a court interpreter. He

Table 3.9. Samuel's empathy map

User: Samuel		Field: Legal	
Place of Origin: Mazatlán Villa de Flores, Oaxaca		Languages: Mazateco, Spanish	
Motivations:	SAY (Quotes and Defining Words)	THINK (Thoughts and Beliefs)	Needs (Usability)
To help his people	"I have seen much injustice done to my Indigenous brothers and sisters for not being able speak Spanish. That is what mainly motivated me to become a professional interpreter." Having only the function of communication bridge "is a challenge that hurts me personally, and I wish I could do more for them."	"Sometimes the Indigenous person confuses the assistance of an interpreter with that of a defense attorney." He sees his function/role as interpreter as a communication bridge but wishes he could do more than that.	To help his community To use different strategies to be able to "deliver a good interpretation"
Challenges:	DO (Actions and Behaviors They Say They Do)	FEEL (Feelings and Emotions)	Needs (Deeper Meaning)
The emotional toll of not being able to help in more ways Untranslatable concepts	"When they ask and say, 'Are you here to help me? Help me,' I tell them that my only function is to be the communication bridge between the two parties." "At the courts, I often encounter legal terms that sometimes you have to paraphrase, you have to pause, and even indicate to the judge to explain the legal terminology in different words."	Needed Helpless Hurt Wounded	To help his community in more ways than only linguistically

sees his role as an interpreter as a communication bridge. Still, he wishes he could do more than that because he sees the injustices that happen against his brothers and sisters for not being able to speak Spanish in Mexico. These injustices are his primary motivation as an interpreter and translator. Although he feels needed, he also feels helpless and hurt that he can't do more for his community. His usability need is to use different strategies to deliver a good interpretation because many legal terms have to be explained as they don't exist in his Indigenous language. His deeper-level need is to help his community.

Abril

Abril is from San Vicente Coatlán, Oaxaca, Mexico (see table 3.10). She speaks Zapoteco del Valle (of the Valley of Oaxaca) and Spanish. She has worked as a court interpreter for almost four years. She is motivated by helping people from her community who are incarcerated. She would like to take more up-to-date courses to learn more strategies to explain concepts she doesn't understand. She feels needed but is also frustrated with the lack of awareness about Indigenous variants. Her

Table 3.10. Abril's empathy map

User: Abril		Field: Legal	
Place of Origin: San Vicente Coatlán, Oaxaca (Mexico)		Languages: Zapoteco del Valley, Spanish	
Motivations:	SAY (Quotes and Defining Words)	THINK (Thoughts and Beliefs)	Needs (Usability)
To help people from her community who are incarcerated	"There are many people, for example, from my community who are in prison and, because they cannot speak Spanish, their processes are delayed and aren't carried out as they should." "There are many who have been released because sometimes they're wrongly imprisoned, or their processes aren't carried out correctly."	"There are terms that sometimes are difficult to explain in our language." She believed that the courts don't respect the rights of Indigenous people because of their lack of knowledge of the Spanish language. "My variant has been requested in the Sierra Sur region, but sometimes they send someone from somewhere else, and this is where they don't get the linguistic variant that is actually needed."	To receive more updated practical training

continued on next page

Table 3.10. Abril's empathy map—*continued*

CHALLENGES:	DO (ACTIONS AND BEHAVIORS THEY SAY THEY DO)	FEEL (FEELINGS AND EMOTIONS)	NEEDS (DEEPER MEANING)
"To take more updated training to continue helping people from my community." Untranslatable concepts: "Many times, for example, they give us a term that sometimes we don't understand. Although we know the theory, there are terms that are difficult to explain in our language."	"I've helped by supporting them in their hearings, and based on this, their processes have moved faster." "I've been doing this for almost 4½ years. I received my credentials in 2014 and have been working since then."	Needed Helpful Frustrated	To help her community Indigenous language variants awareness

usability need is to receive more updated practical training, and her deeper-level needs include helping her community and creating awareness about Indigenous language variants.

Lucas

Lucas is from San Juan Copala, in the municipality of Santiago Juxtlahuaca, Oaxaca, Mexico (see table 3.11). He speaks Triqui de San Juan Copala and Spanish. He became a medical interpreter to help his mom at doctor's appointments when he was only a child. After that, the doctor asked him to help with other Triqui patients because very few people in his community speak Spanish. As an Indigenous teen, he was given a scholarship to study translation and interpretation. He has been an interpreter and translator for the medical and legal fields for over eighteen years. He believes that Native Peoples should be respected in the same way we respect "the French Peoples and the German Peoples." He sees the lack of respect for Indigenous communities and the lack of awareness about Indigenous cultures as the main challenges in his profession. He feels helpful but also invisible and disrespected by the public systems. His needs are at a deeper level, including the need to promote awareness about Indigenous cultures and languages and help his community.

Table 3.11. Lucas's empathy map

USER: Lucas	FIELD: Medical, Legal
PLACE OF ORIGIN: San Juan Copala, of the municipality of Santiago Juxtlahuaca, Oaxaca (Mexico)	LANGUAGES: Triqui de San Juan Copala, Spanish

MOTIVATIONS:	SAY (QUOTES AND DEFINING WORDS)	THINK (THOUGHTS AND BELIEFS)	NEEDS (USABILITY)
He became an interpreter to help his mom in doctor's appointments when he was in middle school.	"Before the conquest, before the invasion, Mexico was also an independent nation, in which each Indigenous community . . . had its own traditions, its own language." "In the community from where I come, all speak the Triqui language. Very few people speak Spanish."	He believes that each community in Mexico is its own nation and that "like we respect other nations, the Mexican Nations must be respected." "Just like the French Peoples and the German Peoples, in Mexico there are Maya Peoples and Aztec Peoples."	NEEDS (Usability)

CHALLENGES:	DO (ACTIONS AND BEHAVIORS THEY SAY THEY DO)	FEEL (FEELINGS AND EMOTIONS)	NEEDS (DEEPER MEANING)
Lack of respect for Indigenous communities Lack of awareness about Indigenous cultures	"I was given an internship to work as a translator for different institutions. But back then there was no payment given for translating; it was more of a social service, and the foundation Telmex gave me a scholarship to continue my studies . . . in the year 2002." "Because I translated for my mom [when in middle school], the doctor called me to translate for other patients who didn't speak Spanish."	Helpful Invisible Disrespected	To promote awareness about the various Indigenous communities and languages in Mexico To help his community

Amanda

Amanda is from Apurímac, Peru (see table 3.12). She speaks Quechua and Spanish. She is an Indigenous educator, interpreter, and translator. Her grandfather died when she was a child, and she had to help her grandmother navigate the legal process. As a young Indigenous girl, she felt intimidated by the protocols of the court system in Peru. She now works in the legal, medical, and educational fields. Amanda believes that there are protocols in the public sector that are intimidating

Table 3.12. Amanda's empathy map

| User: Amanda | Field: Legal, Medical, Educational |
| Place of Origin: Apurímac, Peru | Languages: Quechua, Spanish |

Motivations:	SAY (Quotes and Defining Words)	THINK (Thoughts and Beliefs)	Needs (Usability)
"The need I've seen in my Indigenous communities"	"What has touched my heart, my soul profoundly, is that I'm an Indigenous woman born in an Indigenous community, and I've seen firsthand the need with my own family, my grandmother in particular." She has interpreted for "the National Police, where many community members' rights have been violated."	She believes that there are certain protocols in the public sector that need to be followed that are intimidating for Indigenous people, like taking oaths. She believes that Indigenous rights are often violated.	Court protocols that are more attuned to Indigenous beliefs (or at least sensitive to their ideas)

Challenges:	DO (Actions and Behaviors They Say They Do)	FEEL (Feelings and Emotions)	Needs (Deeper Meaning)
Court protocols Indigenous rights violations	"I'm a teacher by profession. I teach in both my mother tongue and Spanish. That's how I began working toward the visibility and the empowerment of using our mother tongues in public and private spaces." "My grandfather died, and my grandmother was involved in a legal process. . . . My grandmother only spoke Quechua. . . . Every time she went to court, no one understood her . . . I was eight years old when my grandmother told me, 'let's go, you'll accompany me.'" "I had to interpret from my grandmother to the judge and also from the public defendant to my grandmother. Amid this dilemma, I understood that there was an oath involved too, which was intimidating." "They said, 'you have to tell the truth of what she is saying, and only that.'"	Helpful Needed Violated Undervalued Hurt	To advocate for the linguistic rights of her community Visibility for Indigenous beliefs and customs

for Indigenous people. She also believes that Indigenous rights are often violated. Although she feels needed by her community, she also feels undervalued and hurt by the public systems. She feels that her Indigenous rights are constantly being violated. Her usability need is to work in public systems with protocols more attuned to Indigenous beliefs, particularly court protocols. Her deeper-level needs include advocating for her community's linguistic rights and visibility of Indigenous beliefs and customs.

Deconstructing interviews through user empathy maps allows us to learn about the challenges and motivations of the participants, as shown in this chapter. It also gives us a clear view of their thoughts, actions, and feelings to identify surface-level and deeper-level needs. Looking ahead, chapter 4 centers on deconstructing the testimonios shared in the roundtable discussions to identify the specific issues with which Indigenous interpreters and translators grapple. The needs identified in this chapter and the issues defined in chapter 4 are synthesized and analyzed in chapter 5.

4
DEFINING THE ISSUES

TESTIMONIOS AS A UX METHOD

In theory, a design thinking process provides a framework with clear steps delineated in the order we imagine them. In practice, however, as most UX practitioners know, these steps are cyclical because they occur and reoccur as participants bring to light issues and ideas as they articulate them, motivated by the dialogue of and with other participants. Such was the case of the conversations that took place in the roundtable I moderated. Most participants interweaved their personal accounts (which defined the issues) with strategies they have used to counteract injustices toward Indigenous languages (which propelled ideas to solve the issues). I analyze these personal accounts as testimonios.

Like the arc of storytelling, a testimonio builds a narrative, then climaxes at its peak with a desahogo—usually when participants realize that they have said enough (similar to venting), to finally decompress the narrative by working toward a resolution. Even though storytelling does not necessarily aim at advocating for social justice, storytelling becomes the conduit for testimonios to do so (Medina 2018). Through testimonios, participants examine issues within a larger social and cultural context and, therefore, are inspired to find solutions that incorporate their own civic engagement to help produce social change. Testimonios differ from interviews in several ways. The personal narrative in testimonios constructs and reconstructs a lived personal account that connects to a collective experience (Benmayor 2012; Mora Curriao 2007). When sharing testimonios, participants often switch from "I" statements, such as "I experienced . . ." or "When I . . . ," to "we" statements, such as "In my community . . ." or "My family and I . . ." Sometimes there is no clear separation between the individual and the collective because our interactions with others influence our personal experiences (Gonzales, Leon, and Shivers-McNair 2020). Using testimonios in UX research can help us analyze complex matters from a user's *collective* perspective. Further, instead of the researcher asking one question at a time, as in the case of interviews, testimonios draw on prompts to build a narrative

https://doi.org/10.7330/9781646425310.c004

and produce dialogical conversations that seek to understand the perspective of a group of people. It is a research method that can be particularly important when working with underrepresented groups. In the case of a design thinking process, testimonios are even more valuable because participants are dialogically interacting with one another.

For example, after one of the participants in the roundtable, Rosa, talked about her experience as a coordinator of a youth group that provides interpreting services during court hearings in exchange for the university's required social service hours and a stipend, other participants partook in the conversation:

> ANTONIA: I have a question. Where are you all located? Are you in the Public Defender's Office of Mestizos, or do you have a separate space? I'm asking because you talked about the Defender's Office. Are you in a Public Defender's Office?
>
> ROSA: No, actually, the training we get is done in the university facilities. We borrow them. There are many of us, 45, but sometimes other people who are not in the program show up. We've had up to 60 people, and we wouldn't fit in the Public Defender's Office; therefore, we asked the university for their largest classroom.
>
> LUZ: The video cartoons you say are directed to children, are they already translated into their mother tongue?
>
> ROSA: We have only two videos that have been translated because a license is needed, and we only have the rights for two videos. We are barely in the process of applying for other licenses so that more videos can be translated into more languages.
>
> MAGDALENA: And have you all thought of making videos related to your own culture, like myths? Let's say, as a way of contributing to your language and strengthening your own perspectives, your own stories. I don't know if you've thought of that.
>
> ROSA: We haven't thought of that idea, but now that you mention it, yes, we'll consider it.

As seen in this example, the dialogical aspect of testimonios helps untangle complex issues by peeling its layers through dialogue instead of working within the constraints of an interview. Rosa had previously shared that the group she coordinates is committed to promoting Indigenous cultures and languages with children by translating video cartoons. Luz asked Rosa for more details because Luz is involved with a youth group that translates children's books into her P'urhepecha language (I talk more about her project in chapter 6). Antonia is more interested in knowing about the place Rosa's group works from because Antonia also coordinates a group of Indigenous court interpreters in her community in the state of Chihuahua (see table 4.5). Magdalena is a non-Indigenous social anthropologist and human rights educator from Peru who has been involved in Indigenous activism through translation projects (see table 4.4); therefore, she wonders if Rosa and her group can benefit from creating videos using culturally relevant content.

It should be noted that not all participants in this roundtable self-identified as Indigenous, such as in the case of Magdalena, and not all participants shared testimonios. In some cases, I could not record their participation, such as the case of Luz.

My understanding of testimonios incorporates the following characteristics:

- Reconstruct a lived experience through a narrative (storytelling)
- Link a personal narrative to a group's collective experience
- Acquire a reflective dialogical tone
- Call for civic engagement to produce social change
- May involve *desahogarse* as a cathartic act of releasing distressful sentiments (Rivera 2022b)

Testimonios, as Linda Tuhiwai Smith (2012) asserts, is a method often used in Latin American contexts and/or by Latin American scholars to convey the "collective memory" of an oppressed group in order to make sense "of histories, of voices and representation, and of the political narrative of oppression" (145). Because sharing testimonios is an Indigenous practice, using them in research with Indigenous communities prompts us to evaluate the information in Indigenous terms (Rivera Cusicanqui 1987). Clearly, this research method has a lot to offer to UX designers who look for an even deeper understanding of complex issues from the perspectives of Indigenous users using Indigenous research practices.

DIALOGUE AND DESAHOGO AS INDIGENOUS PRACTICES

While dialogue and desahogo are often part of testimonios, each of these practices holds crucial significance. Each of them can and should also be analyzed as independent practices that can help us conduct more meaningful research and create better experiences for Indigenous interpreters.

Dialogue

Throughout the interactions in the roundtable I moderated, dialogue became a tool used by the participants to analyze the complex situations with which they grapple. Dialogue is also a practice that some Indigenous interpreters use during interpreting events. As explained by Pedro during his interview (see chapter 3, table 3.5), it is essential to engage in dialogical conversations with the person for whom they are interpreting because many Western concepts do not exist in Indigenous languages:

> You try to speak with them, not so much to explain certain circumstances but to chat with them so that you can make them understand the circumstances. . . . If an Indigenous person cannot comprehend the concept of "summary proceeding," then you explain, "it is the process or the way." . . . But because I won't be able to explain the concept of "process" either, [then I say,] "it is the way in which they will help you if you accept that you committed a crime." Then again, I won't be able to say "crime," [so I say,] "that you committed an act against someone, a bad action." You tell them, "And if you accept your responsibility, the judge will give you the minimum sentence." You won't be able to explain "minimum" either, so you say, "the least they can."

Dialogue during research interactions, as shown in the last section, and during interpreting events, as demonstrated by Pedro, is what Rivera Cusicanqui (1987) calls an "exercise of collective misalignment" that delinks the parties involved in an oral interaction from their power-powerless relationship. In recent decades, translation and interpreting studies scholarship has also advocated for more dialogical interpreting practices. Wadensjö (2013) argues that an interpreting event that co-constructs meaning between speaker and hearer(s) through dialogue is more equitable because even a stoic poise on behalf of the interpreter can be perceived as a position that favors the public institution. Inghilleri (2012) also understands interpreting as a "living dialogue" that forms and transforms information. Unfortunately, not all Indigenous interpreters can use their dialogical practices because they continue to be expected to remain stoic to demonstrate their *neutrality* to the public institution. And yet, as Slack, Miller, and Doak (1993) argue, interpreters and translators as technical communicators "can never be transparent" because "by virtue of the nature of language, then, [technical communicators] must add, subtract, select, and change meaning" (24). Dialogue not only helps negotiate meaning in oral interactions, it also helps balance power relationships in both research practices and interpreting events.

Desahogo

As narratives, testimonios have an arc that begins with a personal experience that links a collective struggle resulting from a system of oppression. Many testimonio narratives climax in a cathartic, restorative desahogo that yields new possibilities for social change (Rivera 2022b). Whereas the testimonios shared during the roundtable exhibited similar narrative arcs, each individual testimonio was unique in how the elements were displayed. Some centered their testimonios on the collective struggle, some on exploring new possibilities, and some on the political activism that their own civic engagement carries.

Remarkably, the interactions of the group during the roundtable produced a *collective metatestimonio*, where the group built from one conversation to another until reaching a point of a *collective desahogo* that yielded the conscious feeling of "enough is enough" of the group. The group reached the cathartic point of the metatestimonio during the morning of the second day of the event, right after most participants had shared their testimonios and right before we were scheduled to collectively create a list of strategies that could alleviate some of the issues faced by Indigenous interpreters and translators. At that point, Claudia expressed her discontent with the systems and proposed to act upon the written laws that have been created to protect Indigenous linguistic rights (see table 4.6):

> I believe that we, as actors, must demand more from wherever we are, in Chihuahua or Oaxaca. It is by acting and by demanding what has already been written that we'll get results. . . . Let's begin by asking for a dignified wage. . . . There's the General Law of Linguistic Rights; why not use it to demand a budget for the payment of Indigenous interpreters and translators? I believe that we must continue to fight from our own trenches to implement and fulfill what has already been written.

Claudia's desahogo gives us a clear glimpse into the role of Indigenous interpreters and translators beyond authors (Slack, Miller, and Doak 1993) and even beyond advocates (Jones 2016). As technical communicators, Indigenous interpreters and translators are also actors with agency. Bruno Latour's (2005) actor-network theory (ANT) points us in that direction. Latour claims that human and nonhuman actors construct information in an in-flux social network by interacting with one another. While ANT provides a framework for understanding these relationships, it is important to point out that Indigenous communities have always recognized the relationships between human and nonhuman actors. As Shawn Wilson (2008) asserts, an Indigenous ontology is made up of different sets of relationships, and "an object or thing is not as important as one's relationships to it" (73). In this case, Claudia's assertion is a clear example of this material-semiotic relationship. She needs the abstract concept of the *written law* to act, to transform her desahogo into new possibilities that can bring social change.

The idea of an oral desahogo is also present in Indigenous interpreting events, and suppressing it is often the cause of additional emotional stress. During the roundtable, Alejandro explained that when Indigenous interpreters arrive at a court hearing, they are often greeted with a desahogo by the Indigenous person for whom they are to interpret. However, the courts expect Indigenous interpreters to refrain from interacting with the Indigenous victim or defendant beyond the language interpretation. Like Alejandro, others believe that Indigenous court interpreters should be allowed to have a pre-court meeting with the person for whom they are interpreting to better manage the emotional distress:

> Typically, in our Indigenous communities, there is a local authority. Members of this authority speak your language, understand your customs and your traditions because they are from the same community. But when an Indigenous brother comes to a court hearing facilitated by the state government, instead of feeling confident, he feels scared, terrified, and no matter how much he may want to express that he agrees with or doesn't agree with the court or with the judge, he won't do it. Why? Because of the space he's in and the people he's around. First, this Indigenous brother will face a judge, a public defendant, and then the prosecutor. For this reason, we as interpreters, should have the opportunity to have a short briefing with the defendant before the hearing starts in order to let our Indigenous brother know that he is going into a courtroom and that he will face several people, but that he shouldn't be afraid. (Castellanos García et al. 2022, 28)

These situations happen not only in court settings, as Mariana implied during her interview; during interpreting events, medical personnel do not understand emotional desahogos in public spaces (see chapter 3, table 3.2):

> The party for whom you interpret doesn't know your culture. They begin to question, and they begin to offend. Because they don't know our culture, they think that we are acting, but in reality, they don't know our culture.

Regrettably, there are legal consequences for an interpreter who goes beyond interpretation: "They always tell us, 'Do your job right, interpret correctly, or else you are committing fraud'" (Castellanos García et al. 2022, 29).

Allowing a desahogo briefing before an interpreting event provides the interpreter with the context needed to perform a better interpretation and helps release some of the emotional burden upfront. During the roundtable, Alejandro explained that not having a contextual background of the situation provokes tensions that, although cannot be avoided, can at least be mitigated (see table 4.3):

70   DEFINING THE ISSUES

> Imagine the tension when we go to a hearing and the crime is rape, and there is a tabu about these crimes in our Indigenous communities. I cannot interpret for a woman who was raped because the one who raped her was a man, and I am a man. This is an example of the contextual issues we have to consider.

Without a contextual background, it is not possible to know the rhetorical situation of a technical interpretation and/or translation (Castellanos García et al. 2022). As a technical communicator, an interpreter must understand the purpose, the audience, and the context of the situation to localize an interpretation, hence the importance of dialogue and desahogo in Indigenous interpreting events.

MAPPING TESTIMONIOS

To analyze the experiences and the strategies presented by the participants in their testimonios, I crafted a table similar to the user empathy maps proposed by Scott Wible (2020). In this table, which I call a testimonio map, I trace the individual and collective pain points—specific problems experienced by the users (Stanford d.school 2020), the civic engagement activities in which the users have participated, and the outcomes resulting from these civic engagements, all with the purpose of identifying the most significant issues they face. The pain points yielded information that *defines the problems* faced by Indigenous interpreters and translators, and the civic engagement activities and outcomes yielded *ideas* that can be used to address their needs. The Indigenous organization that led the event also invited non-Indigenous participants in order to examine issues from different perspectives. Two non-Indigenous academics joined the conversations in this roundtable. Although twelve participants participated in the roundtable I moderated, not all of them shared testimonios. Therefore, I only analyze the contributions of those who shared their personal experiences. I map the testimonios using a combination of direct and indirect quotations from the conversations recorded.

As with the interviews, I used the sticky note strategy to identify and synthesize themes. While this approach may seem simplistic, it is a process that allows design thinkers to center their attention on the process of the possibilities rather than on the outcome (Wible 2020). My process first involved identifying the pain points discussed by participants in their testimonios. Subsequently, I wrote the themes identified on the pain points on the sticky notes, as shown in figure 4.1. This allowed me to readily move the information to organize it accordingly. To easily

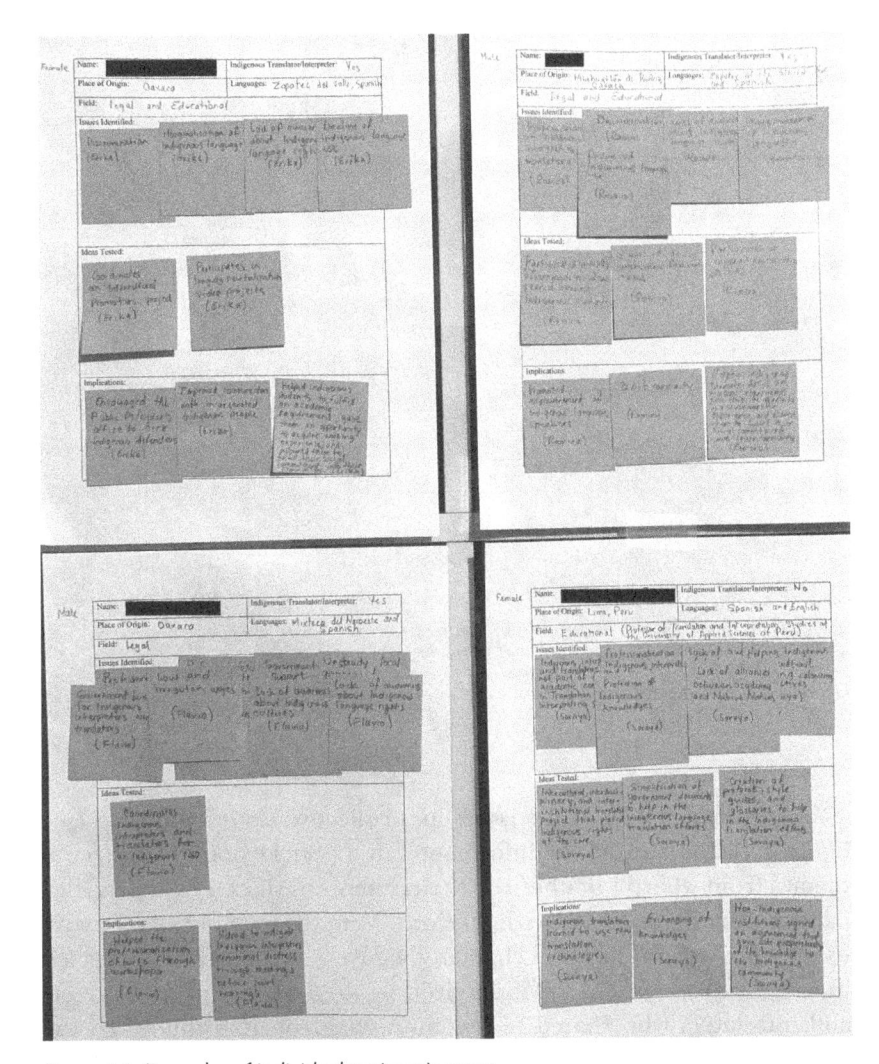

Figure 4.1. Examples of individual testimonio maps

manipulate the information in the sticky notes, I only used the right side of the testimonio maps (Issues, Ideas Tested, and Implications) as there was not enough room to add all the information given in the pain points. Access to a large board would have made this process easier and more fluid. Nonetheless, as shown in the next section, I transcribed the individual testimonio maps in tables that include pain points based on each participant's audio transcription.

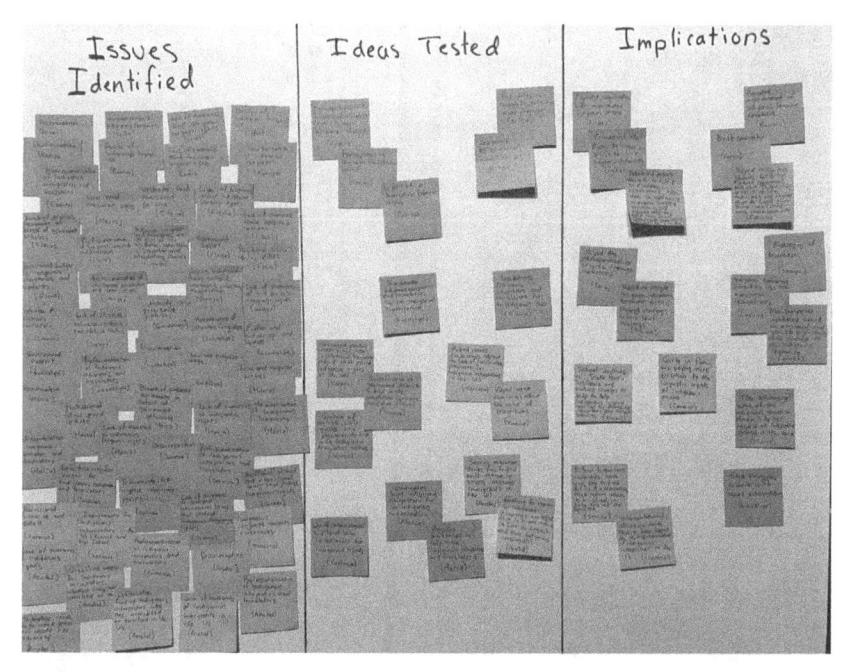

Figure 4.2. Collective testimonio map

Then, I assembled all the issues in a collective testimonio map (see figure 4.2). I organized the information in the sticky notes by collective themes from where I synthesized the testimonios shared by the participants during the roundtable discussions. As in the case of the collective user empathy map shown in chapter 3, figure 4.2 shows how I collected in categories all the information given by the Indigenous interpreters and translators who shared testimonios. Each category mimicked the sections on the right side of the testimonio maps: Issues Identified, Ideas Tested, and Implications of the civic engagement activities in which the participants have taken part. Because I had placed the information on sticky notes (with the name of each participant on each sticky note), this task was easy to do.

As in the interviews, the color of the sticky note bears no relation to how I conducted my research. Once all the information was placed together by category, it was easy to identify themes within each category. And, as in the case of the interviews, because each sticky note had the name of the participant who provided the information, it became easy to identify who said what during the synthesizing process.

TESTIMONIO MAPS

To identify themes in each testimonio map, I organized the ideas of each testimonio by themes and not necessarily in the order of the narrative arc the participants shared it in. I then typed the individual testimonio maps to keep track of each one digitally, as seen in tables 4.1 through 4.9. Whereas most of the pain points and the civic engagements on the left of the testimonio maps are collections of direct quotes arranged by theme, the issues identified, ideas tested, and implications are my interpretations of the participants' direct quotes. As with the user empathy maps, I summarized each testimonio before each testimonio map.

Rosa

Rosa is originally from the State of Oaxaca, Mexico (see table 4.1). She speaks Chinanteco de San Pedro Yolox and Spanish. She coordinates the Intercultural Promoters and Interpreters program at one of the local universities. This initiative prepares Indigenous college students to provide interpreting services during court hearings in exchange for social service hours and a stipend. Rosa believes that Indigenous people involved in the court system in Oaxaca distrust the system and the people they encounter. Statistical information has helped her group create awareness about the current situation of Indigenous languages in the legal field. In her testimonio, she pointed out that the state of Oaxaca houses 178 Indigenous variants and that about 35 percent of the Oaxacan population speaks an Indigenous language. She explained that of the seventy thousand cases in the Public Defender's Office in Oaxaca, thirty thousand concern Indigenous people. Although most public defenders don't speak an Indigenous language, they represent 98 percent of all Indigenous defendants. The issues identified in Rosa's testimonio include discrimination, marginalization of Indigenous languages, lack of awareness about Indigenous language rights, and a rapid decline in the use of Indigenous languages. Her civic engagement activities yielded ideas that included using statistical information to create awareness, engaging Indigenous college students in interpreting services, and dubbing children's videos to promote the use of Indigenous languages. The ideas she proposes have encouraged the Public Defender's Office to hire more Indigenous defenders, have improved communication with incarcerated Indigenous people, and have helped Indigenous students fulfill an academic requirement, gain work experience, and fulfill their social commitment to their Indigenous communities.

Table 4.1. Rosa's testimonio map

User: Rosa	Indigenous: Yes
Place of Origin: State of Oaxaca	Languages: Chinanteco de San Pedro Yolox and Spanish

Field: Legal and educational field. She is the coordinator of the Intercultural Promoters and Interpreters program at one of the local universities, which prepares Indigenous students to provide interpreting services during court hearings in exchange for the university's required social service hours and a stipend.

Pain Points:	Issues Identified:
Lack of Awareness "During the visits to the reintegration centers, we realized that people didn't come to us, not because they didn't want to but because there was no communication bridge because, well, if you can't even defend yourself in Spanish, little less speaking an Indigenous language." "To promote awareness about Indigenous topics in the Public Defender's Office, we had to search statistical information about the people we were servicing." "We know that there are 11 linguistic groups and 68 indigenous languages. In Oaxaca, there are 16 alive languages from where 178 variants derive. As a result, there are in Oaxaca approximately 1,205,000 people who speak an Indigenous language, representing about 35% of the Oaxacan population." "We know that of the 70,000 cases in the Public Defender's Office, 30,000 concern Indigenous people exclusively. We also know that the Public Defender's Office legally represents 98% of Indigenous people who are deprived of their freedom in different reintegration centers in the entire state, which are 13 in total."	Marginalization of Indigenous languages Lack of awareness about Indigenous language rights Decline of Indigenous languages use

Civic Engagement:	Ideas Tested:
"To create awareness about Indigenous topics in the Public Defender's Office, we had to conduct statistical research about the people we served. Today we know that of the 70,000 cases in the Public Defender's Office, 30,000 concern Indigenous people exclusively. We also know that the Public Defender's Office legally represents 98% of Indigenous people who are deprived of their freedom in different reintegration centers in the entire state." "We broadcasted our services through videos, videos that were published on the Facebook page of the Public Defender's Office in 16 different languages, that is 16 variants." "We took interpreters to the correctional hearings to guarantee the linguistic rights of these people." CEPIADET and the Public Defender's Office "provided a theoretical and practical training on topics of Indigenous Peoples human rights, the characteristics of intercultural promoters, translation and interpreting techniques, and topics about the criminal justice system. With this training, interpreters and translators strengthened their abilities, skills, and aptitudes, so they could assist us in helping people." Intercultural promoters and interpreters "also supported us with the translation of texts related to access to justice, and everything that had to do with the rights of incarcerated people and Indigenous people."	Using statistical information to create awareness A project that involves Indigenous college students to interpret at court hearings and to promote awareness about Indigenous cultures. Language revitalization video projects

continued on next page

Table 4.1. Rosa's testimonio map—*continued*

"In the next training, there's going to be a project where they have to go to their communities. We are proposing that they show cartoons, for example, Bugs Bunny from Warner Bros. I have an example in my cell phone where the audio has been replaced with a Zapotec language. This is one of the ideas we have for young interpreters and translators to continue revitalizing [Indigenous languages] even more. . . . We have only two videos that have been translated because a license is needed, and we only have the license for two videos. We are in the process of applying for other licenses so that more videos can be translated to more languages."

Outcomes:	Implications:
"Thanks to this visibility, the Public Defender's Office was motivated to prioritize hiring bilingual defenders so that there could be better communication among incarcerated Indigenous people." "This, as I mentioned, betters the interaction among incarcerated people and allow us to fulfill our [school's] social service requirement one hundred percent." "We have now 45 young people taking our trainings." "These young people in our program are paid 3,600 Mexican pesos each month."	Encouraged the Public Defender's Office to hire Indigenous defenders Improved communication with incarcerated Indigenous people Helped Indigenous students fulfill an academic requirement, gave them an opportunity to acquire working experience, and allowed them to fulfill their social commitment with their Indigenous communities

Carlos

Carlos is from Miahuatlán de Porfirio Díaz, Oaxaca (see table 4.2). He speaks Zapoteco de la Sierra Sur and Spanish. He is an engineering student who participates in various initiatives as an intercultural promoter and interpreter. In his testimonio, he pointed out that although most older adults in his town only speak Zapoteco, children are losing the language because their parents teach them Spanish in an effort to save them from embarrassment and discrimination. He also pointed out that most men emigrate to the United States. He would like to receive more training to address the many challenges that overwhelm Indigenous interpreters. The issues identified in his testimonio include the professionalization of Indigenous interpreters and translators, discrimination, lack of awareness about Indigenous language rights, marginalization of Indigenous languages, and the decline of Indigenous language use. His civic

Table 4.2. Carlos's testimonio map

User: Carlos	Indigenous: Yes
Place of Origin: Miahuatlán de Porfirio Díaz, Oaxaca	Languages: Zapotec of the Sierra Sur and Spanish

Field: Legal and educational field. He is a university student and a member of the Intercultural Promoters and Interpreters program. He studies engineering and development in business innovation.

Pain Points:	Issues Identified:
Discrimination "Most of the town [elders], about 90%, speak an Indigenous language, Zapoteco. Adults suffer discrimination when they move to the city or work here and there." "When I was at school, I was embarrassed to speak it." "I got interested because the majority of the men in my town leave for the East or West Coast of the United States." *Lack of Awareness about Indigenous Language* "There are many people who have encountered violence and discrimination and have been incarcerated. And these are people who cannot speak Spanish perfectly, and their situation becomes more difficult." *Decline of Indigenous Languages* He teaches Zapoteco to the kids in his community. "These are kids who are learning Spanish and are letting go of their Indigenous language." "Nowadays in my community the majority are losing [the language], more so the kids, because their parents are teaching them Spanish, because of the same reason, parents don't want their kids to suffer embarrassments or discrimination by other kids whose first language is Spanish when they arrive at the city." *Professionalization* "We're seeking training in all aspects of legal proceedings, specifically to address all the challenges that may overwhelm the interpreter."	Professionalization of Indigenous interpreters and translators Discrimination Lack of awareness about Indigenous language rights Decline of the use of Indigenous languages Marginalization of Indigenous languages

Civic Engagement:	Ideas Tested:
"One day, I met a professor who also spoke an Indigenous language, Zapoteco, but a different variant." She said, "I have a friend who is launching a campaign, the 'Choose Wisely' campaign, which was launched by SEDESOH [Secretaría de Desarrollo Social y Humano] here in Oaxaca. She asked if I could help translate my language so that the campaign could be launched statewide, and people could really listen to what we are trying to communicate." "Since then, I became interested in translating and in knowing my language because I was embarrassed before." "When the campaign was over, I learned about the program through the Public Defender's Office in Oaxaca. I saw the information on Facebook, and I said, 'I'll register in this program and let's see what happens.' This is the reason why now I am one of the students in the Indigenous translators group." "The Secretaría de Pueblos Indígenas y Afroamericanos launched a scholarship for young Indigenous people who are attending university. I also benefitted from this program. In this program, we must create a project to rescue and strengthen our Indigenous languages. My proposal consisted of visiting my community on Saturdays, for at least two to three hours, to teach kids who are in fifth and sixth grade. . . . I talk about my Indigenous language and teach them to speak it, and to write it so that they can continue to preserve the language."	Participated in government and NGO initiatives geared toward Indigenous students Is part of a program of intercultural promoters Participates in language revitalization projects

continued on next page

Table 4.2. Carlos's testimonio map—*continued*

Outcomes:	Implications:
"Now I am proud of my Zapoteco language, from the Sierra Sur of Oaxaca, of Miahuatlán de Porfirio Díaz." "My experience with the Public Defender's Office has been very interesting because there are lots of us. We initiated an enormous group, and it is nice to see that we all care about our Indigenous languages."	Promoted empowerment among Indigenous language speakers Built community Helped Indigenous students fulfill an academic requirement, acquire work experience, and fulfill their social commitment to their communities

engagement activities yielded ideas that include participating in initiatives geared toward students offered by NGOs, the government, and universities to promote cultural awareness and revitalize Indigenous languages. These ideas have instilled empowerment among Indigenous language speakers, built community, and helped Indigenous students fulfill an academic requirement while allowing them to acquire work experience and fulfill their social commitment to their Indigenous communities.

Alejandro

Alejandro is from Miramar, Oaxaca, Mexico (see table 4.3). He speaks Mixteco del Noroeste and Spanish. He coordinates a group of Indigenous interpreters and translators. He works almost exclusively in the court systems. He believes that the main issues faced by Indigenous interpreters are the lack of consistent professionalization programs that address the specific needs of Indigenous interpreters and translators, lack of professional follow-up and updates that calibrate and assess the efficiency of interpreting techniques, lack of consistent government support (because the support and policies change when the government administration changes), lack of government budget for Indigenous interpreters and translators in the court systems, low and irregular wages for Indigenous interpreters and translators, breach of payment agreements on behalf of government officials, lack of awareness about Indigenous cultures, and lack of awareness about Indigenous language rights. As a coordinator of Indigenous interpreters and translators, he helps organize workshops that support interpreters and translators. He also helps mitigate Indigenous interpreters' emotional distress through meetings before court hearings.

Table 4.3. Alejandro's testimonio map

User: Alejandro	Indigenous: Yes
Place of Origin: Miramar, municipality of Santa María Yucuhiti, district of Tlaxiaco, Oaxaca	Languages: Mixteco del Noroeste and Spanish

Field: Legal. He is the coordinator of a group of interpreters and translators of Indigenous languages.

Pain Points:	Issues Identified:
Government Policies "The Public Defender's Office of Oaxaca belongs to the state government. They have their own defenders, and some of them speak Indigenous languages. They have their own office in Ciudad Judicial, but they are defenders who belong to the state government. . . . It is very important what the Defender's Office is doing. Not all incoming governments have that perspective, and sometimes they abandon Indigenous topics." *Professionalization* "We can send interpreters to a court hearing, but if this person is not translating well in the context, or if an interpreter doesn't know the technique or the topic, especially now that all is done orally, then it creates confusion. . . . And now that the federal courts are paying interpreters, we worry about the quality of the interpretation." "In Mexico, there are certified, accredited, and practical interpreters. The majority of the interpreters are either accredited or practical. We only have one certified interpreter. These are our main issues, because in Oaxaca, as far as I know, there are only nine interpreters certified. Imagine! From the 176 variants spoken in Oaxaca, only nine interpreters are fully certified. Judges are required to look for certified interpreters first. Then, if there are none, they look for an accredited one, then for a practical interpreter." "We all believe that becoming a certified interpreter is important, but we have to see the other side of the issue. The only ones who certify at a national level are INALI [Instituto Nacional de Lenguas Indígenas] and CONOCER [Consejo Nacional de Normalización y Certificaión de Competencias Laborales]. They certified interpreters in the years 2008, 2010, 2012, and 2014. We asked where the interpreters that were certified in the year 2008 are. There is a registry, the PANITLI, which is the Padrón Nacional de Intérpretes y Traductores en Lenguas Indígenas, but that doesn't guarantee you a job or payment." "We conducted a certification research in 2010, and we found out that after being certified, interpreters never went to a court hearing. Maybe it was because their linguistic variant wasn't required, or maybe another reason, but the issue is that no one followed up to ensure they continued to receive training and updates. I believe that it is not only important to certify Indigenous interpreters and translators but also there has to be a follow-up in the form of training, alignment courses, and assessments on behalf of the institution that provided the certification." "Who wouldn't want a certification? But not if they leave you alone in the journey without continued training and assessment. Practicums are needed, but you also have to know the interpreting techniques, like paraphrasing and emotional control."	Professionalization of Indigenous interpreters and translators Professional follow-up and update Government support Inconsistent local government policies Government budget for Indigenous interpreters and translators Breach of payment agreements on behalf of government officials Lack of awareness about Indigenous cultures Lack of awareness about Indigenous language rights Low and irregular wages for Indigenous interpreters and translators

continued on next page

Table 4.3. Alejandro's testimonio map—*continued*

"Imagine the tension, when we go to a hearing and the crime is rape, and there is a tabu about these crimes in our Indigenous communities. I cannot interpret for a woman who was raped because the one who raped her was a man, and I am a man. This is an example of the contextual issues we have to consider."

"We have hard data that proves that many times practical interpreters do a much better job than the certified ones. However, we must create awareness about this with public officials because they always want certified interpreters."

Wages

"The federal court in Oaxaca pays 1,800 Mexican pesos per hearing for up to 3 hours, and sometimes they pay for travel expenses.... The common courts pay only 400 Mexican pesos per hearing and that includes travel expenses.... Imagine the expenses of an interpreter who comes from Coixtlahuaca to Salina Cruz. From Coixtlahuaca to Oaxaca City you have to spend about 200 pesos. From there, you have to take a car to Salina Cruz, which is another 120 pesos. That's a total of 320 pesos, but that is just one way, and that is only transportation, and it doesn't include food and lodging, then it is not enough.... And if the interpreter who made the commitment to attend doesn't show up, then this interpreter is fined."

"If they call you to tell you, 'you have a hearing tomorrow,' and you are in the Mixteca region, and you have to be in the Istmo region, who guarantees your expenses? Interpreters don't go to hearings when no one takes care of their expenses."

"Sometimes the courts don't want to pay interpreters at the end of the hearing, nor do they want to give them the paper that proves that they attended the hearing.... Then, they owe money to many interpreters. They owe to interpreters from [our organization] and others who work independently, who have told me that the courts owe them money since 2017 or 2018."

"We have also done translations for the electoral court of the state of Oaxaca on narratives based on the Indigenous system of 'customs and traditions.' In this case, the translator doesn't have to travel, you send the work via e-mail. The INPI pays 200 Mexican pesos per sheet. The INALI pays between 350 and 400 pesos per sheet, but the federal electoral court pays 800 pesos per sheet.... There are translators who know about these different rates and always ask to be paid 800 Mexican pesos."

"The government is supposed to guarantee the payment of interpreters and translators. But sometimes we receive family members from someone detained in the courts, or someone involved in a court hearing, who say, 'It's urgent that I have an interpreter for my hearing, so that the process continues, and, if needed, I'll pay for the interpreter.' 'But this is not your responsibility,' I say, 'the government is supposed to look for the interpreter and to pay for the service.'"

Awareness of Indigenous Language Rights

"There are many laws in Mexico that address Indigenous language rights and the right to have an interpreter in legal procedures, like the Federal Law of Linguistic Rights of Indigenous Peoples, articles 9 and 10, the second article of the Mexican Constitution, the National Code of Criminal Procedures, and the agreement 169 of the OIT [Organización Internacional del Trabajo, International Labor Organization], but sometimes the reality is another thing."

continued on next page

Table 4.3. Alejandro's testimonio map—*continued*

CIVIC ENGAGEMENT:	IDEAS TESTED:
"I'm an interpreter, and I'm also a coordinator of interpreters and translators of Indigenous languages. . . . I work mostly with that organization, where there are needs for interpreters and translators in the common courts, federal courts, and electoral courts, as well as for translations and interpretations in the courts in the United States. I'm the one who receives all the requests to facilitate communication and to assign an interpreter."	Coordinates Indigenous interpreters and translators for a Mexican NGO

OUTCOMES:	IMPLICATIONS:
"We'll work with the federal court at the end of the year to start a series of workshops to help in the professionalization of interpreters. . . . We cannot certify though." "Before we send an interpreter to a court hearing for the first time, we make an appointment with the interpreter to go over the background of the hearing to help with the emotional distress."	Helped the professionalization efforts through workshops Helped mitigate Indigenous interpreters' emotional distress through briefings before court hearings

Magdalena

Magdalena is a social anthropologist and a human rights professor in Lima, Peru (see table 4.4). She is also a leader of an international law organization. She is not an Indigenous woman and doesn't speak an Indigenous language. Still, she has worked with the Indigenous community of Achuar del Pastaza in Peru to create bilingual birth, death, and marriage certificates for this Indigenous community. In her testimonio, she explained that Indigenous translators and interpreters are not part of the academic conversations in translation and interpreting studies. She also pointed out other issues that include the lack of effective professionalization programs for Indigenous interpreters and translators, lack of awareness about Indigenous language rights, helping Indigenous Peoples without imposing colonizing perspectives, lack of protection for Indigenous knowledges, and too few alliances between academia and Native Nations. She proposes more intercultural, interdisciplinary, and interinstitutional projects that place Indigenous rights at the core. To work on projects like hers (described in chapter 6), she proposed simplifying government documents and creating protocols, style guides, and glossaries to help in the Indigenous language translation efforts. The project in which she participated created an alliance between the Achuar Indigenous community and academia, helped Indigenous translators to become familiar with new translation technologies, and fostered an exchange of knowledges between Indigenous and non-Indigenous translators, all while acknowledging through a written contract that the Achuar Indigenous community was the sole proprietor of the knowledge translated in the academic institution for which she works.

Table 4.4. Magdalena's testimonio map

User: Magdalena	Indigenous: No
Place of Origin: Lima, Peru	Languages: Spanish and English

Field: Educational. She is a social anthropologist and a human rights professor in Peru. She is also a leader of an international law organization.

Pain Points:	Issues Identified:
Awareness "Achuar del Pastaza is an Amazonian Indigenous community of approximately 5,000 people, all of whom speak Achuar. . . . This town asked the government to provide them with the public service of bilingual certificates of birth, death, and marriage, which are called 'vital events.' These certificates are made by the Registro Nacional de Identificación y Estado Civil [RENIEC], you must have an equivalent here, which is the institution in charge of giving your DNI [Documentación Nacional de Identidad (National ID)]. If someone dies, there is a certificate. And also, if someone is born, so that this person can acquire a DNI." *Professionalization* The university's Translation and Interpreting program got involved because "for this service to be comprehensive, they had to translate about 60 different documents. A document for a name change, a document because you make a mistake, many documents. Then, they did not have resources anymore. They only had two people translating." There were no Achuar translators, so the project organizers sought out the help of "two Achuar educators, who were not interpreters or translators. They were bilingual educators but took on the role of professional translators." "Because UPC was used to working with foreign languages, they did not want to go to the Achuar community. . . . For the first time, two Achuar people came to UPC." "Imagine! The academia, which only teaches Chinese, French, German, and Italian, suddenly brings Indigenous educators with a different language, a different worldview, a different cosmovision." "In the academy, no one really knew how to translate the Achuar language. Who was going to say if it was well done or not?" *Protection of Indigenous Knowledges* "With an intercultural focus and an emphasis on rights, what was most important, I think, is that there was trust. . . . The emphasis on Indigenous rights was really important; in other words, this translation was not intended to perpetuate colonization. This translation was a project for the community to be able to exercise their rights in their own language." *Alliances between Academia and Native Nations* "I believe that we should promote alliances between academia and Native nations, but always within an Indigenous rights lens that prevents the university from taking their knowledge."	Indigenous translators and interpreters are not part of the academic conversations in translation and interpreting studies Professionalization of Indigenous interpreters and translators Lack of awareness about Indigenous language rights Helping Indigenous Peoples without imposing colonizing perspectives Protection of Indigenous knowledges Lack of alliances between academia and Native Nations

continued on next page

Table 4.4. Magdalena's testimonio map—*continued*

CIVIC ENGAGEMENT:	IDEAS TESTED:
"I am here because of a project [I was involved in] focused on providing access to justice for speakers of Indigenous languages. In addition, I'm also a member of [an international organization] which has programs focused on training and education, and in which we are now also starting to focus on issues of translation." "That's how the project began between the UPC, Universidad de Perú de Ciencias Aplicadas, and the FENAP, which is the Federación of the Nacionalidad Achuar del Perú, and the Instituto Internacional de Derecho y Sociedad [IIDS]." "We signed an interinstitutional and intercultural agreement because Achuar educators came to work hand in hand with professors at the university"	Intercultural, interdisciplinary, and interinstitutional translation project that placed Indigenous rights at the core Simplification of government documents to aid Indigenous language translation efforts Creation of protocols, style guides, and glossaries to help Indigenous language translation efforts

OUTCOMES:	IMPLICATIONS:
"This was an interdisciplinary experience because we saw the intercultural lens, the linguistic lens, the translation lens, and the Indigenous rights lens. Each institution contributed something. For example, the intercultural lens came from the academy, from my class; the linguistic lens, from the linguists at the university; the translation lens, from the translators; and the Indigenous rights lens came from Instituto Internacional de Derecho y Sociedad [IIDS]." "I want to highlight the details that happened from the linguistic lens. It was very interesting. For example, the certificate said, 'By means of this letter, we present to the registrar . . .' with a bombastic language that had the Achuar community all dizzy. What did we do with the pool of linguists? We simplified the document. What does this mean? This means that we simply focused on what was said, who said it, and to whom. That simple." "At the university, no one really knew how to translate the Achuar language. Who was going to say if it was well done or not? Therefore, from the linguistic lens, we only agreed on the methodologies and a bit on the translation process. And because there were certificates that had been translated before, then the terminology was validated for the next documents. We also created protocols. We were looking for quality in the translation. Because there were two translators, one corrected the translation of the other. The protocols consisted of identifying how a particular term was translated and why. Then, glossaries were created because, as I stated before, there were terms like identity, birth, death, and so on. And some of these terms did not exist in the Achuar language. For example, 'go to the registrar's office' did not exist in Achuar, then, translating became more complex. Stylesheets were also created to ensure the quality in the translation of certificates. If they added an upper-case letter in one place, then they added it in the whole document. If they added a comma after something, then they had to maintain the same format throughout the document. We wrote notes on the stylesheets and the protocols."	Creation of alliances between Indigenous communities and academia Indigenous translators learned to use new translation technology Exchanging of knowledges A non-Indigenous institution signed an agreement that gave sole proprietorship of the knowledge translated to the Indigenous community

continued on next page

Table 4.4 Magdalena's testimonio map—*continued*

"They used a program, software, to help the Achuar translators with time management and style, as well as other things. We developed a system in Memsource, which segments the text, allows translators to know how much is left in the translation, and incorporates terms that have already been translated into the text. With this software, they were able to complete the translation."

"With an intercultural focus and emphasis on rights, what was most important, I think, is that we built trust. . . . The emphasis on Indigenous rights was really important; in other words, this translation was not intended to perpetuate colonization. This translation was a project for the community to be able to exercise their rights in their own language, which was also really gratifying in terms of strategic partnerships between academia; the RENIEC, which is a state institution; FENAP, which is an Indigenous organization; an Indigenous community; and also our civil organization through the International Institute [IIDS]."

"Imagine! The academia, which only teaches Chinese, French, German, and Italian, suddenly brings Indigenous educators with a different language, a different worldview, and a different cosmovision."

"We signed an agreement with an Indigenous rights lens, an agreement where we clearly stated that the university would not be able to appropriate their language. What was translated belonged to the community. We created very clear protocols so that the university would not say, 'This is mine, and I'm going to do whatever I want with the Achuar language.' No. We protected Indigenous language rights."

"There was also an exchange of knowledges because, for example, the linguists would say, 'How are we going to ask them for their password?' because the online resources needed passwords, 'How is this going to work? Surely they don't know about passwords.' And it turned out that the Achuar educators had two terms for passwords. Why? Because they had worked in Ecuador, and the policies are more advanced over there, so they had two terms. That's an example of how the cultures clashed and how academia was challenged."

Antonia

Antonia is from Chihuahua, Mexico (see table 4.5). She speaks Tarahumara and Spanish. She is the coordinator of a nonprofit organization that aims to train Tarahumara court interpreters. She has seen how public officials sometimes look for people on the streets to interpret for them instead of hiring (and paying) a certified interpreter. The main issues she has faced include low and irregular wages for Indigenous interpreters and translators, professionalization of Indigenous interpreters and translators, lack of professional follow-up and update, discrimination, lack of government support, inconsistent local government policies that change with every new government administration, and the marginalization of Indigenous languages. Although her organization has been able to provide training for a group of forty-four interpreters, she still struggles to work with court administrators.

Table 4.5. Antonia's testimonio map

USER: Antonia	INDIGENOUS: Yes
PLACE OF ORIGIN: Chihuahua, Mexico	LANGUAGES: Tarahumara and Spanish

FIELD: Legal. She works for a state government agency and coordinates a nonprofit organization of Tarahumara court interpreters.

PAIN POINTS:	ISSUES IDENTIFIED:
Lack of Government Support "Those who work for the justice departments don't care. They look for people [to interpret] on the streets, even if they aren't certified. Sometimes they even look inside the CERESO prison [for someone to interpret for free]." "We have laws, but the governments shut their eyes and ears and overlook the issues. Sometimes this happens because they are temporary governments. Those who come in, arrive without knowing how to continue the work or the commitment that others made." *Professionalization* "I think that we should get together to stop injustice and for that we have to certify ourselves. . . . We should be able to decide who should be certified." "Sometimes there is favoritism. People certify someone only because the person is a good friend, but then that person is not going to work well as an interpreter." "If those who train us don't have time to follow up with interpreters, to find out where they are and if they are working, then they should ask for help. . . . They should really want to help us." *Wages* "There are many issues, like irregular wages. They aren't the same everywhere. In Chihuahua, they pay 200 Mexican pesos per hearing, and someone else said they pay 400 in Oaxaca, and then 1,800 in the federal court. Why does it vary like this? . . . We should tell them, 'If you don't come to an agreement, we can manage interpreters for you. Just give as a space and we take care of your problem.'" *Discrimination* "We have heard about human rights and about discrimination for a long time, and we are still talking about the same issues with no solutions. Sometimes there's even discrimination among ourselves. What should we expect from those who aren't of our race?"	ISSUES IDENTIFIED Low and irregular wages for Indigenous interpreters and translators Professionalization of Indigenous interpreters and translators Professional follow-up and update Discrimination Lack of government support Inconsistent local government policies Marginalization of Indigenous languages

CIVIC ENGAGEMENT:	IDEAS TESTED:
"I have a nonprofit organization. . . . I will have forty-four interpreters in the second group joining this organization. I'm the coordinator, the one who coordinates those working at the hearings."	Coordinates a group of Indigenous court interpreters

OUTCOMES:	IMPLICATIONS:
"So far, we haven't been really successful because sometimes the court administrators don't want to give us the opportunity to work. They think they are the only ones who can participate [in the court hearings]."	Still struggles to work with court administrators

Claudia

Claudia is originally from Chiapas but now lives in Mexico City (see table 4.6). She speaks Tzeltal de los Altos de Chiapas and Spanish. She coordinates Indigenous interpreters and translators for an Indigenous NGO in Mexico City. In her testimonio, she drew attention to the low visibility Indigenous immigrant communities have in Mexico City. The main issues that she identified in her profession are low and irregular wages for Indigenous interpreters and translators, professionalization, lack of professional follow-up and training updates, lack of awareness of Indigenous rights, lack of awareness of Indigenous language rights, discrimination, marginalization of Indigenous languages, and breach of payment agreements on behalf of government officials. As the coordinator of interpreters and translators in the organization where she works, she coordinates workshops to help prepare Indigenous translators and interpreters. Although she sees the many issues faced in her profession, she acknowledges that the advocacy work of one Indigenous group in Mexico City has helped all Indigenous groups in the area.

Table 4.6. Claudia's testimonio map

User: Claudia	Indigenous: Yes
Place of Origin: Originally from Chiapas but has lived in Mexico City for over twenty years	Languages: Tzeltal from the Altos de Chiapas and Spanish

Field: Legal. She coordinates a group of Indigenous interpreters and translators for an Indigenous NGO based in Mexico City.

Pain Points:	Issues Identified:
Professionalization "INALI and CONOCER give you a paper that says that you are proficient in your language, but that doesn't mean that I am apt to interpret in that language." "The other day I went to the office of INALI and I run into a Mixtec colleague from Oaxaca who was visiting Mexico City, and we began to talk, 'What are you doing here?' 'I came to pick up the certificate I earned four years ago and to find out what's happening,' my colleague said, 'because they haven't called me, I haven't worked as an interpreter not even once. I'm certified now, but they haven't called me, so I'm here to find out what's happening and to leave my contact information.'" *Wages / Discrimination* "The wages for Indigenous interpreters aren't defined. Are we supposed to receive less than an interpreter of a foreign language? Because interpreters of foreign languages get paid very well."	Low and irregular wages for Indigenous interpreters and translators Professionalization of Indigenous interpreters and translators Lack of professional follow-up and training updates Lack of awareness of Indigenous rights Lack of awareness of Indigenous language rights Discrimination Marginalization of Indigenous languages

continued on next page

Table 4.6. Claudia's testimonio map—*continued*

"The problem is that you go and translate or interpret, but they don't pay you until three or four months down the road. In fact, they owe us since 2015. Then, our colleagues ask, 'When? By when?' And we as nonprofit have to insist so that our colleagues get paid for their time because ultimately the Indigenous interpreters end up doing the work for the authorities."	Breach of payment agreements on behalf of government officials Local government policies

Discrimination / Local Government Policies

"I don't know if you all heard the speech by the Judge of Morelos. There I saw the difference between the foreign and the Indigenous, and the discrimination. He said that an economic ruling was asked, and that surveyor was paid 24,000 Mexican pesos. Then, he said that an anthropological ruling was asked, and that surveyor was paid 12,000 Mexican pesos. What is happening?"

Awareness

"There is a constant struggle. But since the Triqui Indigenous movement in Mexico City took place, the public policies in Mexico City have paid more attention to the Indigenous migrant population . . . because when there are Indigenous migrants, the local government washes its hands because they think that's not their problem."

"I believe that we, as actors, must demand more from wherever we are, in Chihuahua or in Oaxaca. It is by acting and by demanding what has already been written that we'll get results. . . . Let's begin by asking for dignified wages. . . . There's the General Law of Linguistic Rights, why not use it to demand a budget for the payment of Indigenous interpreters and translators? I believe that we must continue to fight from our own trenches to implement and fulfil what has already been written."

CIVIC ENGAGEMENT:	IDEAS TESTED:
"I am currently part of an organization of translators and interpreters of Indigenous languages. I'm part of the advisory board and my job right now is to funnel interpreters and translators into the federal sector, especially with the Council of the Federal Administration. I work on local affairs, which often require the presence of a translator, particularly in Mexico City. Sometimes, we also have to go outside of Mexico City. That's mainly what we focus on, in addition to also giving classes, workshops, and workshops to train translators."	Coordinates Indigenous court interpreters for an Indigenous organization Provides workshops to help train Indigenous translators and interpreters

OUTCOMES:	IMPLICATIONS:
"I think that in Mexico City a door for us has been opened, but this has been the result of a struggle."	The advocacy work of one Indigenous group in Mexico City has helped all Indigenous groups in the area.

Lourdes

Lourdes is originally from Peru but now lives in Belgium (see table 4.7). She is a professor of translation, interpreting, and intercultural studies. Although she is not an Indigenous woman and does not speak an Indigenous language, she advocates for Indigenous rights through her work as an academic and international consultant. She believes that the main issues faced by Indigenous translators and interpreters include the lack of awareness of local and international laws that protect Indigenous rights and Indigenous language rights, the low and irregular wages for Indigenous interpreters and translators, professionalization of Indigenous interpreters and translators, discrimination, and the colonizing perspectives still imposed by governments. She uses international and local laws to advocate for Indigenous rights, which have helped Peru draw more attention to the linguistic rights of Indigenous people.

Table 4.7. Lourdes's testimonio map

User: Lourdes	Indigenous: No
Place of Origin: Peru but lives in Belgium	Languages: Spanish, English, Dutch
Field: Education and government initiatives. She is a professor of translation, interpreting, and intercultural studies. She is also an international consultant.	

Pain Points:	Issues Identified:
Professionalization "Indeed, the State is the one who offers the accreditation [for Indigenous interpreters and translators]." "The certification operates on two levels: the accreditation of taking a course and the administrative process that comes with completing a series of practicums in a public institution. And finally, the registry. . . . Any public institution, the Health Ministry or the Justice Ministry, can access this registry and see in the list the language for which an interpretation or translation is needed." "There are intensive courses given every semester, and after a certain number of hours, that Indigenous translator and interpreter is accredited. Then, there's a national registry, but that isn't enough because Indigenous interpreters have to be professionalized. In other words, Indigenous interpreters haven't necessarily completed a professional degree; they aren't necessarily healthcare experts, they aren't necessarily court process servers, and they aren't necessarily linguists, although there are some. This doesn't mean that there are none. Most likely they have been an Indigenous leader within their community, appointed by the community, and sent to Lima, the central authority, to take this course." "About the training given by the Ministry of Culture, some people say that it is intensive, but there are cases, for example, where someone hasn't received a postsecondary education that allows this person to understand a Western system regarding judicial administration to comprehend what is an advanced process, commutation of sentence, or simply matters of intercultural healthcare, like what it means to be given anesthesia. For some Indigenous women, to be under anesthesia is to die." *Wages* "A problem transpires though, what happens if the interpreter comes from a Peruvian Amazonian Achuar community and has to take first a boat of various hours of river journey to later arrive in Lima, and then take a plane [to the final destination]? Who pays for the travel expenses? Who pays for the transportation? The reimbursement isn't clear. A plane ticket costs 150 dollars minimum."	Lack of awareness of local and international laws that protect Indigenous rights Lack of awareness of local and international laws that protect Indigenous language rights Low and irregular wages for Indigenous interpreters and translators Professionalization of Indigenous interpreters and translators Government policies Discrimination

continued on next page

Table 4.7. Lourdes's testimonio map—*continued*

Government Policies

"The State doesn't qualify them as Indigenous interpreters, no, for the State, they are 'interpreters of native languages.' However, some interpreters want to be recognized as Indigenous interpreters."

"In cases of prior consultation, for example, Indigenous interpreters can be seen as people who help the process of the State and not as someone who is really conveying the needs of the Indigenous community."

Awareness of Indigenous Language Rights/Local Government Policies

"In Peru, the certification and accreditation of interpreters of Indigenous languages was created following the Convention No. 169 [of 1989] by the International Labor Organization [ILO], and it draws on the Native Peoples' fundamental right to express in their own language when discussing matters with the judicial authorities of the State. Convention No. 169 requires that Native Peoples be consulted in different matters. To make provision for this, a law of prior consultation, in their own language, was established. But then, the State says, 'from the needs of the State, and not of the Native Peoples.'"

"In Peru, the law establishes the right, and I quote, 'to enjoy and to have access to translation means . . . that guarantee the enactment of rights in all sectors.' . . . But this is only one step."

Civic Engagement:	Ideas Tested:
"I work at the University of Antwerp, and I'm also an international consultant. What brings me to Oaxaca today is the opportunity to share time and experiences with all of you, and more than anything, to learn from you and get to know more about commonalities between Mexico and Peru with regards to Indigenous translators and interpreters."	Uses international and local laws to advocate for Indigenous rights

Outcomes:	Implications:
In her testimonio, she discussed a court case in Peru that ruled in favor of an Indigenous woman speaker of Quechua. A previous court had made her sign a document that only allowed her to prepare food outdoors certain hours of the day. The judge found out that the woman had signed with her fingerprint, which meant that she did not speak Spanish, signing a paper in Spanish without understanding it. "The court said, 'This is not a case of the right to work, it is a case of the right to not being discriminated against for using a particular language.' . . . And stated that this was a discrimination case because there was no equity. The court also stated that there is not only an individual right to use Indigenous languages but also a collective right because there is a majority of Indigenous speakers, and the State is required to recognize that language as official and to provide the necessary means. . . . But this was only one step."	Courts in Peru are paying more attention to Indigenous discrimination and to Indigenous linguistic rights.

Julia

Julia is from California (see table 4.8). She speaks English and Spanish. In her testimonio, she emphasized that in the United States, official certification for Indigenous languages of Latin America doesn't exist. The main issues she pointed out are the lack of an effective and consistent professionalization system for Indigenous interpreters and translators, requirements for Indigenous interpreters in the United States (English, Spanish, an Indigenous language, and high school), professional follow-up and updates, and awareness about Indigenous language variants. She has helped by creating awareness about the lack of certification programs for Indigenous interpreters and translators in the United States and raising awareness about the value of practicums. Her advocacy efforts have helped a few Indigenous interpreters pass the difficult examinations in the United States, which require them to be fluent in three languages, English, Spanish, and their Indigenous language. She has also helped coordinate interinstitutional alliances with Mexico to assist in the professionalization of Indigenous interpreters and translators in the United States.

Table 4.8. Julia's testimonio map

User: Julia	Indigenous: Yes
Place of Origin: California	Languages: Spanish and English
Field: Legal and medical	

Pain Points:	Issues Identified:
Professionalization "In the United States, there's no certification for Indigenous languages of Mexico. There are legal and healthcare certifications at the national level, but only for the three predominant languages: Spanish, Chinese, and another language." *Professionalization in the Medical Field* "There are three [languages in which interpreters can be certified] in the medical field, which are done through two independent agencies. . . . They offer what's called 'CoreCHI' [Certification for Healthcare Interpreters]. You can still get certified [in other languages], but you have to know English and Spanish. You can be an Indigenous person who speaks an Indigenous language and still get a national certification in the healthcare industry. Still, you are required to know English and Spanish. What's happening in our communities is that most speakers of Indigenous languages only speak Spanish." The healthcare accreditation consists of "attending a course with a certain number of hours, which covers standards, ethics, and protocols based on the guidelines of the agency CCHI [Certification Commission for Healthcare Interpreters]. If anyone wants to become certified in that field, they have to pay 500.00 dollars. I believe that there's only a handful of interpreters in the state of California who have this national certification in the healthcare industry."	Professionalization of Indigenous interpreters and translators Requirements for Indigenous interpreters in the United States (fluent in three languages and having a high school education) Professional follow-up and update Awareness of Indigenous language variants

continued on next page

Table 4.8. Julia's testimonio map—*continued*

"We want to establish networks for evaluation since sometimes there aren't even three people who speak the same language, and we don't want to be the only ones who can certify ourselves. We want to learn more about this process for all of those who speak Mixteco in their own language variant because nobody else will be able [to certify]. Other people do understand the language, but sometimes they don't understand it well enough."

"There are different perspectives on what a certification really means, because sometimes those who are certified, despite them having a certificate, do not understand things the same as those who have done a practicum."

Professionalization in the Legal Field

"In the legal field, there are three categories for interpreters: certified, professionally qualified, and skilled."

"At the legal and national certification level, there is also a prerequisite, which is to have at least high school. This is a criterium that many of our interpreters do not meet. And in the US, this has to be met in English. You can take your high school courses here in Mexico, but you also have to take the GED, a high school equivalency test for adults, which is administered in English, so that you can meet the education requirements to practice in the United States."

Civic Engagement:	Ideas Tested:
"What I've been doing over the last six years is to let agencies know that there is no certification program in Indigenous languages, which many agencies are not aware of. Therefore, when I look at someone who has a practicum, they already have lived experience that helps them cover those services, or they are what we call 'qualified,' which is very similar to an accreditation."	Helped create awareness about the lack of certification programs for Indigenous interpreters and translators in the United States Helped create awareness about the value of practicums

Outcomes:	Implications:
"I have been working with a group of Indigenous women to help them pass their CoreCHI examination in English and Spanish. It took them almost two and a half years to be able to pass the exam." "We've been in conversation with INALI, for at least a year and a half, to be able to take . . . the process of certification they have already structured, so that we can replicate and use it as a standard so that we don't reinvent the wheel. To use it, but to also localize it to our own communities."	A few Indigenous interpreters have been able to pass difficult examinations that require them to be fluent in three languages Interinstitutional alliances with Mexico have helped the professionalization of Indigenous interpreters and translators in the United States

Valeria

Valeria is originally from Michoacán but now lives in Santa María, California (see table 4.9). Although she does not speak an Indigenous language, she works with the Mixteco migrant community in Santa María. She works as an intermediary between the Mixteco migrant community and the school districts. A major part of her work consists of finding Indigenous interpreters that can help communicate between parents of Indigenous students and the schools. The issues she identified include difficulties finding Indigenous interpreters who are accredited or certified, professionalization of Indigenous interpreters and translators, dignified wages for Indigenous interpreters whether they are certified or not, lack of awareness about Indigenous immigrants in the United States, discrimination, and lack of awareness about Indigenous language rights. In her testimonio, she also pointed out the need to use technology more effectively in advocacy efforts. Some of the ideas that sprout from her civic engagement include looking for resources for Indigenous interpreters to help them become certified in English and Spanish and sharing immigrant stories during meetings to build community among Indigenous immigrants in the United States. Her advocacy work has prompted school districts in Santa María, California, to make changes to help Indigenous migrant families by providing interpreters and other services.

Table 4.9. Valeria's testimonio map

User: Valeria	Indigenous: Yes
Place of Origin: Originally from Michoacán but now lives in Santa María, California	Languages: English and Spanish

Field: Educational. Although she doesn't speak Mixteco, she works as an intermediary between the Mixteco migrant community and the school districts in Santa María, California.

Pain Points:	Issues Identified:
Professionalization "Because I have to work with communities that are neither fluent in English nor Spanish, I have to recruit any interpreter available, whether they are accredited or certified or have experience. Either way, I am going to do what I can to communicate with Indigenous families." "Once Indigenous parents find out that we have an interpreter who can help, there's more demand, and right now we don't have enough people who can interpret, and we don't have enough training available in our city either."	Technology needs to be used more effectively for advocacy Difficulties finding Indigenous interpreters who are accredited or certified

continued on next page

Table 4.9. Valeria's testimonio map—*continued*

"We recruit interpreters over 18 years old, and we try not to recruit students because we don't put students younger than 18 years old in traumatic situations. It is very difficult for students to try to interpret for mom and dad a painful situation in which they were involved." *Wages* "I work with interpreters who want to help others from the heart and who are willing to work voluntarily, those who say, 'I'll go whether they pay me or not.' Then, I look for training and ways to ensure that the district pays for the professional development of the interpreter. Once this interpreter has received the appropriate training, the district hires this person as a full-time interpreter." "The school district has the funds, and the key person in the school district and I work together to ensure that the interpreter gets paid." *Awareness about Indigenous Migrant Communities* "The most important thing for us, for the interpreter, for the district liaison, and for me, is to ensure that Indigenous immigrants are heard in the school districts." *Discrimination* "We also want to ensure that their language is sustained, and that the students over there don't feel embarrassed to speak their languages." *Use of Technology* "We have had the idea of creating videos to express ourselves orally, but for one reason or another, even though in the United States there are lots of resources, we haven't been able to make a video like that."	Professionalization of Indigenous interpreters and translators Dignified wages for Indigenous interpreters whether they are certified or not Lack of awareness about Indigenous immigrants in the United States Discrimination Lack of awareness about Indigenous language rights

CIVIC ENGAGEMENT:	IDEAS TESTED:
"It's been priceless to work with trilingual interpreters who have to know a form of Mixteco, as well as Spanish and English. So, I'm here at this conference to learn more from you all, so that I can take more information over there to work more effectively with interpreters and to help people over there in the US." "I coordinated a program where these families shared what their life is like in the city of Santa María, describing their migration trajectory."	Locating resources for Indigenous interpreters has helped them to become certified interpreters in Spanish, English, and their Indigenous language Sharing migration stories has helped build community among Indigenous immigrants in the United States

OUTCOMES:	IMPLICATIONS:
"Through our work, we were able to share information with the school districts, and they're now making changes to help Mixteco families, particularly students who only speak Mixteco."	School districts in Santa María, California, are making changes to help Indigenous migrant families by providing interpreters and other services

Testimonios help identify elements of the user experience that other methods ignore. Through their dialogical narrative, testimonios link a personal experience with a collective struggle while fostering civic engagement. From the narrative of testimonios stems *dialogue* and *desahogo* as Indigenous practices that can be used not only in interpreting events but also in UX research, as this chapter shows. In the next chapter, I synthesize both the interviews deconstructed in chapter 3 and the testimonios reviewed in this chapter by identifying themes that draw attention to the collective needs and the collective issues faced by Indigenous professionals in the field of translation and interpretation.

5
SYNTHESIZING NEEDS AND ISSUES

As an interdisciplinary field that addresses communicating information effectively through technology or about technical or specialized topics (Society for Technical Communication 2023), technical communication involves *translating* technical information for nontechnical audiences. Far too often, however, this definition hides a connotation that understands technical communication primarily as communicating *written/visual* information through or about *digital* technologies. Angela Haas (2012) asserts that "people tend to conflate new media with definitions of technology, heralding as technologies only the emerging ones and relegating to tools or crafts the older and more ubiquitous technologies" (288). Thus, technical communication concepts are rarely applied to those who work with technical information *orally* in physical spaces. Reaffirming that technical communication can be, and often is, conducted orally in nondigital spaces helps us situate the discussions in this chapter because although Indigenous interpreters and translators can, and often do, conduct their work in written form through digital technologies, their work is largely carried out orally and in the presence of the users for whom they perform their work.

Grounding the work of Indigenous interpreters and translators as technical communication enables us to better understand their motivations, challenges, feelings, self-perception, needs, and issues through their own lenses. It also allows us to clearly see how and why Indigenous interpreters and translators have always been concerned with matters of social justice, because the spaces where their technical and professional communication occurs are imbued with injustice. To cite an example, specialized terms and jargon used in the legal, medical, and educational fields are already biased as they favor the hierarchical Western systems for which they are created. The biased jargon used in these spaces also reinforces ideologies of homogeneity and *blanquitud*. Bolívar Echeverría (2010) situates blanquitud, or whiteness, in relation to modernity and capitalism. For Echeverría, blanquitud is "the full set of visible traits that accompany productivity, from the physical appearance of the body

https://doi.org/10.7330/9781646425310.c005

and its environment, clean and ordered, to the language properties, the discrete positivity of the attitude and the gaze and the restraint and composure of the gestures and movements" (59). To survive in today's modern and capitalist societies, as Echeverría points out, a dangerously homogenous appearance and attitude of blanquitud is expected.

Technical and professional spaces, in particular, demand that we speak a Western language—with a Western accent, that we know Western technical skills, and that we display a "professional" demeanor that restrains emotions. Not surprisingly, then, Rebecca Walton, Kristen R. Moore, and Natasha N. Jones (2019) assert that the field of technical communication is "complicit in injustice" (8). Indigenous interpreters and translators, therefore, are largely motivated by a commitment to help their Indigenous communities and a desire to work toward redressing the injustices Western systems produce. It is because of these injustices that they feel needed by their communities but neglected by the public systems in which they work. And it is injustice caused by discrimination and lack of awareness about Indigenous matters that drives Indigenous interpreters and translators to see their work as acts of activism.

MOTIVATIONS, CHALLENGES, FEELINGS, AND SELF-PERCEPTION

Technical and professional communicators face different challenges and thus find motivation in different factors. Some are motivated by professional advancement, some by earning a decent wage, and some by helping people. Although these factors also motivate Indigenous interpreters and translators, their strong communal identities, as shown in this section, prompt Indigenous professionals to predominantly find motivation in helping their communities navigate situations of injustice. Since not all the participants in this study contributed to both interviews and testimonios, it is important to provide a separate demographic and contextual background for each group.

As seen in table 5.1, the participants who contributed to the interviews included seven females and five males, three from Peru, two from the United States, and seven from Mexico. Ten participants in this group work in the legal field, four in the medical field, four in the educational field, and two for government initiatives. The linguistic skills of the participants in this group included one speaker of Tzeltal de los Altos de Chiapas (from the highlands of the Mexican state of Chiapas), one speaker of Zapoteco del Valle (of the Valley), one speaker of Nahuatl, one speaker of Zapoteco del Sur (of the South), one of

Table 5.1. Baseline characteristics of participants who contributed to the interviews

Baseline Characteristics	n	%
GENDER		
Female	7	58
Male	5	42
COUNTRY		
Mexico	7	58
Peru	3	25
United States	2	17
FIELD[a]		
Educational	4	33
Government Initiatives	2	17
Legal	10	83
Medical	4	33
LANGUAGES[b]		
English	2	17
Jaqaru	1	8
Mazateco	1	8
Mixteco Alto	1	8
Mixteco Bajo	2	17
Mixteco del Noroeste	1	8
Nahuatl	1	8
Quechua de Cusco	2	17
Spanish	12	100
Triqui de San Juan Copala	1	8
Tzeltal	1	8
Zapoteco del Sur	1	8
Zapoteco del Valle	1	8

a. *Some participants work in two or more fields.*
b. *All participants speak two or more languages.*

Mixteco del Noroeste (of the Northeast), one of Mazateco, one of Triqui de San Juan Copala, two speakers of Mixteco Bajo (from the lowlands), one of Mixteco Alto (from the highlands), two speakers of Quechua de Cusco (from Peru), one speaker of Jaqaru (from Peru), two speakers of English, and twelve speakers of Spanish. All participants in this group self-identify as Indigenous and work as interpreters and/or translators. Empathy maps and biographical sketches of each participant can be found in chapter 3.

Motivations

For technical and professional communicators in and outside of academia, engaging with communities is not an uncommon practice. TPC scholarship and practices have supported community-based projects

Motivations
(12 Total Participants)

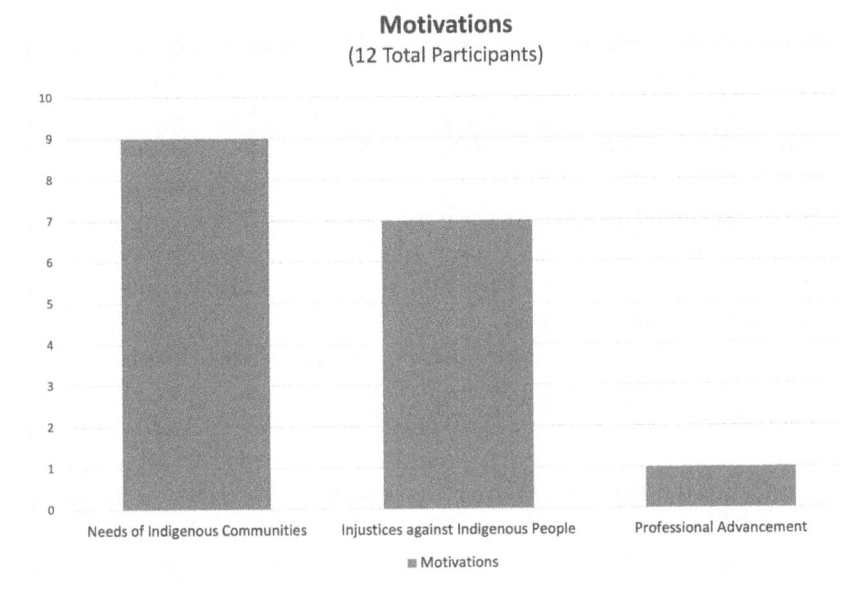

Figure 5.1. Motivations of Indigenous interpreters and translators

that help meet the communication needs of underrepresented communities by designing or re-designing technical documents (Durá 2015; Durá, Gonzales, and Solis 2019; Gonzales et al. 2022; Rose et al. 2017; St.Amant 2020). While these important contributions promote more inclusive practices, insufficient attention has been given to what motivates Indigenous individuals in their roles as technical communicators and users. Indigenous interpreters and translators are not necessarily motivated by the same factors that motivate most of us. Their work is largely motivated by the needs of their Indigenous communities, and their communal contributions are not circumscribed by short-term projects but are understood as lifelong commitments, as this study shows.

Most participants in this study stated that the main motivation in their profession is the linguistic needs of their Indigenous communities (see figure 5.1). Nine of the twelve participants interviewed affirmed that they became involved in this profession as a result of the necessity to help either their own parents or community members. For example, during her interview, Mariana stated that she became an interpreter because she saw the need in her community: "I started to contribute with my voice, the most important communication bridge to them, whether in the medical field or the educational field. And I began to discover that the need was much bigger. That motivated me to start in this profession as an interpreter." Like Mariana, Natalia stated that her main motivation is to help her community and "to watch over [her] people, in all its cultural aspects." Seven of the twelve participants said they were motivated by the injustices they saw occurring against Indigenous people. Abril, for instance, expressed in her interview that she became an interpreter to help her community navigate Western legal processes: "There are many people, for example, from my community who are in prison and, because they cannot speak Spanish, their processes are delayed and aren't carried out as they should be. This motivated me to join this profession."

During the interviews, only one of the twelve participants, Pedro, stated that he became involved in the court system at first to advance in his profession, but then he was asked to help interpret because he was the only person who spoke the Zapotec language needed at that moment: "I introduced myself to this court, Juzgado de Control y Tribunal de Enjuiciamiento, to conduct my internships and social service. After they found out that I spoke this Indigenous language, I was asked to help translate for defendants and victims. This is how I got to work with this court, and now I collaborate with them." Clearly, the commitment that most Indigenous interpreters and translators have toward their profession is inherently connected to the needs of their communities.

Challenges

We can no longer ignore that Western hegemony has excluded Indigenous cultures, knowledges, and practices from professional settings, so much so that the distant cultures of Western Europe are seen as domestic while Native cultures of the Americas are seen as foreign and extraneous. Iris Young (1990) believes that this "cultural imperialism" is a face of oppression as present as violence, powerlessness, marginalization, and exploitation. Drawing on the feminist theories of María Lugones and Elizabeth Spelman (1983), Iris Young (1990) argues that one dominant group's claim to universality causes cultural imperialism, turning other cultures inferior, at best, and at worst, invisible: "The dominant group reinforces its position by bringing the other groups under the measure of its dominant norms. Consequently, the difference of women from men, American Indians or Africans from Europeans, Jews from Christians, homosexuals from heterosexuals, workers from professionals, becomes reconstructed largely as deviance and inferiority" (59). The Indigenous participants in this study put cultural imperialism in milder terms and called it *lack of awareness.*

Challenges
(12 Total Participants)

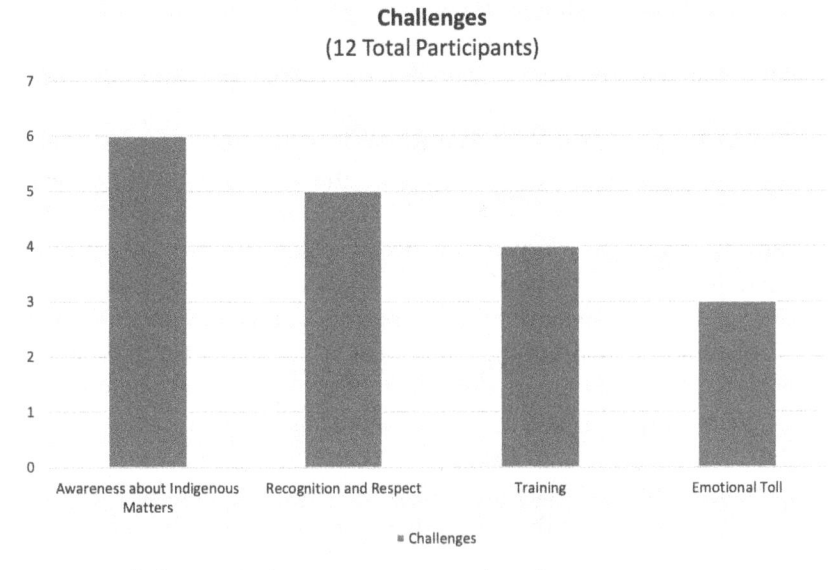

Figure 5.2. Challenges of Indigenous interpreters and translators

As seen in figure 5.2, participants identified lack of awareness about Indigenous matters as their primary challenge. Six participants believe that public officials and staff working for public institutions know very little about the multidimensional cultures of Indigenous communities and about the linguistic rights that protect them, as Claudia pointed out during her interview: "As an interpreter, I've noticed the lack of sensibility on behalf of the authorities, which are the ones who are involved in the rendering of justice. I have noticed that there is still much ignorance about Indigenous Peoples issues, and there is a lack of awareness, perhaps even ignorance, about the current laws that benefit Indigenous Peoples." Unfortunately, laws that protect Indigenous rights, particularly linguistic rights, are not widely known within Indigenous communities either (public officials do not normally spend time with Indigenous communities disseminating Indigenous laws). Gabriela clearly explained why her role as an interpreter is critical not only to make a communication message accessible for two users from different cultures but also to create awareness about Indigenous rights: "[The job of an interpreter] is very important because there is a lot of need in the Indigenous communities. Many Indigenous people don't know their rights, and many wouldn't be where they are if they knew their rights, if they stood up and said 'enough.' And they can do that with an interpreter."

Another major challenge related to lack of awareness is the scarce information about Indigenous language variants. In Mexico, for example, there are eleven linguistic families, from which sixty-eight linguistic groups stem. From these sixty-eight groups, nonetheless, 364 variants originate. Because of the marked differences between variants, differences that speakers of two variants from the same group find difficult to understand, the Instituto Nacional de Lenguas Indigenas (INALI) in Mexico has urged everyone for decades to treat each variant as an independent language (INALI 2008). To put this linguistic awareness challenge into perspective, there are forty-three Mayan variants in Mexico alone (INALI 2008). This poses a challenge for Indigenous interpreters because they are often requested to interpret for a linguistic group without specifying the exact variant. This, in turn, causes a delay in the process in which an Indigenous person is involved. For instance, Abril pointed out that her variant "has been requested in the Sierra Sur Region, but sometimes they send someone from somewhere else, and this is where they don't get the linguistic variant that is actually needed." Natalia explained during her interview that it is imperative for people to acknowledge and respect Indigenous language variants: "I would like people to know that there are many languages in each country, Indigenous languages, and that there are also many variants, and that these variants differ from one another, and that these variants should be respected when people translate."

Like simplifying the many linguistic differences of Indigenous groups, lack of professional recognition and respect for the work of Indigenous interpreters and translators are effects of privileging Western systems and languages. Privilege, as Walton, Moore, and Jones (2019) have argued, hinders the view of those who are at the center of society, making it difficult "for the most privileged individuals and groups to notice others' oppressions (or how they are complicit in others' oppressions)" (84). Nonetheless, privilege does not exist as a simple dichotomy between centered groups and marginalized groups. Privilege intersects, intertwines, and interacts with individuals and groups in complex ways. Individuals of one group and, therefore, their practices and languages can be at the margins when interacting within one society while being at the center when interacting within another, such as the case of Mestizes in the United States and Mestizes in Latin America, to name one example. And these complex interactions change over time, as Walton, Moore, and Jones point out.

Because Indigenous languages do not possess the same value as European languages, Luis asserted in his interview, one of the greatest challenges for Indigenous interpreters is "the recognition of the work of the interpreter." For instance, numerous initiatives in Peru to create awareness about Indigenous languages have emerged directly from Indigenous bilingual educators. Yet, these efforts rarely render job recognition or job status in a monetary sense, as Natalia explains:

> The greatest support in human resources is the educators who teach bilingual education. There is a bilingual education program, and there are people committed, adults and young, who identify themselves as Jaqaru, who are Jaqaru, and want what's best for them. Economically, no. We don't have resources. We are committed people. Maybe it is the collective work among us, among us and for us. But in this [economic] aspect, we're not yet [successful]. In academic and government institutions, our work needs to be recognized as any other professional working as a translator or interpreter.

Other challenges that participants identified are training and the emotional toll of interpreting in a space that is zealously guarded by a Western cosmovision. In the case of training, Indigenous interpreters do not receive updated preparation, especially in the technical concepts of the fields in which they work. They know the theoretical basis of their job as translators and interpreters, but they feel that they need more technical preparation that is specific to each field, as Abril stated: "Many times, for example, they give us a term that sometimes we don't understand. Although we know the theory, there are terms that sometimes are difficult to explain in our language." Similarly, Mariana indicated that there is a need for more training to help Indigenous interpreters manage the emotional toll: "One of the daily challenges has been to prepare myself emotionally because we don't know the situation that the job will bring day by day." In her research with interpreters from Alicante, Spain, Dalila Niño Moral (2008) shows that interpreters feel under substantial emotional tension as they navigate through the different contexts of their work while controlling the strong emotions triggered by the tragedies they witness because they think that they are expected to *only* mediate language. This is an issue that, like Mariana implied, could be addressed not only by providing training to interpreters on strategies to manage emotions but also by creating awareness among public institutions about the unrealistic perception and expectation of the role of interpreters as they clearly mediate a lot more than language (Biernacka 2008). Although what I mention here relates to the linguistic challenges of Indigenous interpreters, these challenges derive from more systemic issues that I will discuss later in this chapter.

Feelings

Identifying how Indigenous interpreters and translators feel is essential because understanding the users' point of view at that level "foreground[s] insights about the emotional depth and breadth of a person's experiences rather than only their material needs" (Wible 2020, 413). While users' feelings are always crucial in the participatory design process, identifying feelings is essential when working with marginalized communities negatively impacted by colonialism. Although I am identifying feelings based on the interviews, testimonios can be more useful than interviews in this area because testimonios afford dialogues and desahogos that can reach deeper-level analyses.

Figure 5.3 shows that all Indigenous interpreters who participated in the interviews feel helpful and needed by their communities. However, they also feel overlooked and neglected by the public systems with which they come in contact. Alejandro noted during his interview that "there are federal institutions that are already offering dignified wages, but there are still statewide institutions and other authorities that do not want to offer dignified wages for Indigenous interpreters." Five participants stated that they feel aggravated and disrespected by public officials and/or by staff members of public institutions. Amanda, for instance, said during her interview that when her grandfather passed away, her grandmother did not receive the help she needed from the courts in Peru: "My grandfather died, and my grandmother was involved in a legal process in the Corte Superior de Justicia of Apuriac. My grandmother only spoke Quechua and did not know Spanish at all. Every time she went to court, no one understood her. She just stood there day after day." Mariana also indicated that because most people working in public institutions do not know her Mixtec culture, "they begin to ask questions, and they begin to offend. Because they don't know our culture, they think that we are acting, but in reality, they don't know our culture."

During the interviews, Samuel expressed that he actually feels hurt and helpless because he can only help by interpreting and wishes he could do more: "Sometimes the Indigenous person confuses the assistance of an interpreter with that of a defense attorney. It is when they ask and say, 'Are you here to help me? Help me,' that I tell them that my only function is to be the bridge of communication between the two parties. It is here where I cannot do more for my Indigenous brothers and sisters. This is a challenge that hurts me personally, and I wish I could do more for them." Victoria, moreover, feels afraid that she could hurt her Indigenous community by interpreting or translating for the government because if she conducts work that "only benefits the government,

How Do Indigenous Interpreters & Translators Feel?
(12 Total Participants)

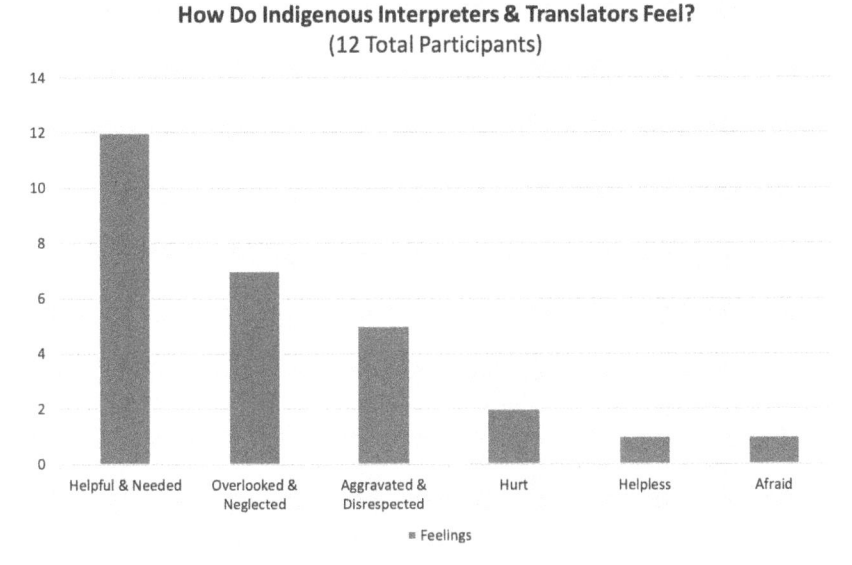

Feeling	Count
Helpful & Needed	12
Overlooked & Neglected	7
Aggravated & Disrespected	5
Hurt	2
Helpless	<1
Afraid	<1

Figure 5.3. Feelings of Indigenous interpreters and translators

that violates the rights and laws, then [she] would be selling out human beings, which are [her] people." Although many participants expressed feeling neglected and overlooked by public institutions, all feel that they are serving the needs of their own communities, hence the importance of continuing their work despite the circumstances.

Self-Perception of Role

Public institutions differ in their perceptions of the role of interpreters (and sometimes of translators too) from machine-like conduits to more participatory roles, often depending on the field in which they work (Angelelli 2004; Gentile 2014). Translation and interpreting studies scholars believe that these discrepancies in public institutions influence interpreters' self-perception of their role (Niño Moral 2008; Gentile 2014). There are discrepancies and confusion about the *ideal* and the *real* role of interpreters, as they are often told to translate "word for word," provoking an "inter-sender conflict" about the role required by one sender and the role expected by another (Gentile 2014, 200). While this "inter-sender conflict" is real during an interpreting event, most Indigenous interpreters have a clear understanding of their role, as seen in figure 5.4.

Eight of the twelve participants understand their role as an act of activism. For example, Gabriela commented that many Indigenous

How Do Indigenous Interpreters & Translators See Their Role?
(12 Total Participants)

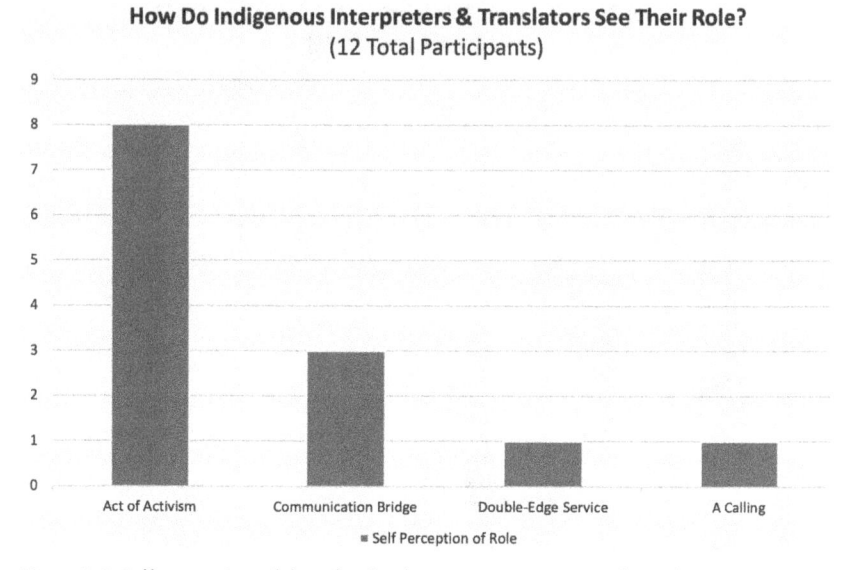

Act of Activism	8
Communication Bridge	3
Double-Edge Service	1
A Calling	1

Figure 5.4. Self-perception of the role of Indigenous interpreters and translators

people are incarcerated because they do not know their rights. She believes that "with an interpreter, they can know their rights and that an interpreter can help with their needs." Similarly, Natalia explained that her experience as the first translator in her Jaqaru language "has been very rewarding because it gave visibility to [her] people and gave it a certain status, without losing the symbolic meaning of the Jaqaru People."

Three of the participants see their role as a communication bridge. For instance, although Samuel wishes he could do more for his Indigenous community, he sees his role as a communication bridge: "Sometimes the Indigenous person confuses the assistance of an interpreter with that of a defense attorney. When they ask and say, 'Are you here to help me? Help me,' I tell them that my only function is to be the bridge of communication between the two parties." Claudia understands her role as "a calling." She commented that she became an interpreter and translator because she noticed the many problems "in matters of justice" that her Tzeltal community face in Mexico City. "It is here when I realized that being an interpreter was my calling," she stated. Victoria, however, believes that interpreting and translating for Indigenous communities is a "double-edged delicate matter." Although Indigenous interpreters and translators often help their communities, they can also hurt Indigenous people if their work only benefits the other party, as Victoria pointed out. "In the case of prior consultation, it is a double-edged work. I could be Native, Indigenous Andean, speaking my Native language, but if my work only benefits the government, which violates the rights and laws, then I would be selling out human beings, which are my people. Then this becomes a double-edged delicate matter."

Whereas Indigenous interpreters and translators' self-perception of their role is strongly linked to purpose and commitment, public institutions perceive their role very differently. The perceptions that public institutions have of the role of interpreters and translators are connected to the perceived status of that profession (Gentile 2014), which is a factor associated with the place of employment, which is associated with the job (in the same way we associate doctors working at hospitals and teachers working at schools). And because most interpreters and translators, whether Indigenous or not, are employed on a contract basis, their professional status in the eyes of public institutions creates inconsistent perceptions of their roles (Gentile 2014). Paola Gentile (2014) suggests that the nature of the job as a service profession that intends to address a need—a communication need—is another important factor that affects the perceptions of the role of interpreters and translators. Although this study does not fully clarify the profession's role as seen through the lens of public institutions, it proves that there are evident conflicting perceptions of the profession between Western and Indigenous worldviews. Western views understand the role of Indigenous interpreters and translators through a *status* lens while Indigenous views see this role through a *purpose* lens. Clearly, Indigenous interpreters and translators believe that their role comes with agency, which they understand as advocacy, which they see as a critical need as discussed in the next section.

NEEDS IDENTIFIED THROUGH INTERVIEWS

Participants who contributed to the interviews identified five critical needs: to contribute to their Indigenous communities, to advocate for Indigenous rights and Indigenous visibility, to be recognized and valued as an Indigenous interpreter and/or translator, to receive updated training, and to be able to use Indigenous practices—such as dialogue and desahogo—when interpreting. Whereas eight of the twelve interviewees identified contributions to their communities and Indigenous advocacy as critical needs, only three participants identified training and using Indigenous practices in translating and interpreting events as key needs (see figure 5.5). Four participants identified recognition as a major need. The differences between the number of participants who recognized one need over another do not make one need less important than

Needs
(12 Total Participants)

Need	Critical Needs
Contribute to Community	8
Advocate for Indigenous Rights & Visibility	8
Be Recognized & Respected	4
Receive Updated Training	3
Be Able to Use Indigenous Interpreting Practices	3

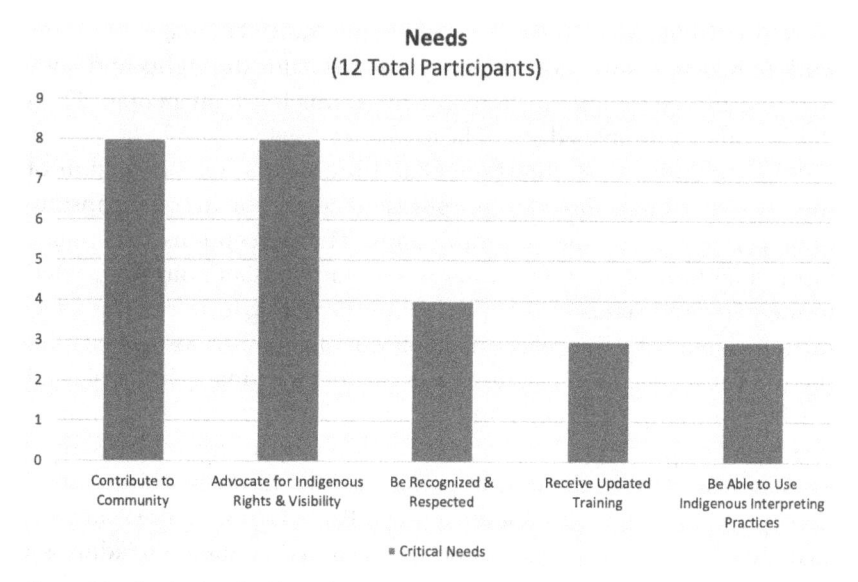

Figure 5.5. Needs identified by Indigenous interpreters and translators

the other. What these results highlight is the high value that participants place on collective needs over their individual needs.

Contribution to Community

While interpreting and translating is a service profession by nature because it meets a communication need, Indigenous interpreting and translation goes beyond addressing a communication necessity. It addresses a *community* need. Linda Tuhiwai Smith (2012) asserts that "affirming connectedness" to a community through "spiritual relationships and community well-being" is part of the identity of Indigenous people (149, 150). A clear illustration of this is Rigoberta Menchú's (1984) narration of when her community held a meeting to initiate her, at the age of ten, into adult life: "Then [my parents] made me repeat the promises my parents had made for me when I was born; when I was accepted into the community; when they said I belonged to the community and would have to serve it when I grew up" (49). In a similar way, most Indigenous interpreters and translators identified contributing to their communities as a critical need, as Luis explained: "At first, I studied to become a teacher of Indigenous education at the elementary level in the area of Nahuatl, but then I was interested in supporting my community because of the necessity of interpreters and translators, primarily because of the injustices experienced by Indigenous communities and by Indigenous people."

Like Luis, Mariana stated that she uses her voice as a tool to contribute to her community. This motivates her to continue working as an Indigenous interpreter, mainly because she discovered that "the need was much bigger" than she initially thought: "I saw the need that the community had, that because of the language barrier, many injustices were committed. And I began contributing with my voice." Amanda also expressed that helping her community through her profession is why she is in this profession: "Because of the need I've seen in my Indigenous community, I participate as a translator, as an interpreter, in the legal, medical, and educational fields." Using their language skills to contribute to the well-being of their Indigenous communities is an essential practice of Indigenous professionals.

Cana Uluak Itchuaqiyaq and Breeanne Matheson (2021) argue that when working with Indigenous groups or adopting Indigenous decolonial knowledges, we too must consider how our scholarship will directly benefit Indigenous Peoples and communities. Doing so ensures that our scholarship acknowledges and respects Indigenous viewpoints, which, as shown here, are situated around Indigenous communities' advocacy.

Advocacy for Indigenous Rights and Indigenous Visibility

Another critical need that also relates to the collective responsibility that the participants have with their communities is the need to advocate for Indigenous rights, sovereignty, and visibility. Indigenous people demand their inherent right "to determine their own communicative needs and desires in the pursuit of self-determination" (Lyons 2000, 462). Indigenous interpreters and translators, like Indigenous scholars (Rivera Cusicanqui 1987; Lyons 2000; Smith 2012; Tuck and Yang 2012), call for their linguistic sovereignty by advocating for their oral practices and their languages.

During her interview, for instance, Amanda explained that part of her work has been to advocate for the visibility of Indigenous languages: "I am a teacher by profession. I teach in both my mother tongue and Spanish. That's how I began working toward the visibility and the empowerment of the importance of using our mother tongues in public and private spaces." Similarly, as the first official translator and interpreter in the Jaqaru language in Peru, Natalia explained that being a translator and interpreter has been very rewarding for her because her work has given "visibility to [her] people." In their role as technical communicators, advocating for linguistic sovereignty, both written and oral, is a critical need for Indigenous interpreters and translators.

And yet advocacy is a concept that is perceived differently by the fields on which this study is grounded. On the one hand, advocacy is still largely perceived as an unethical practice by TIS programs and by public institutions—because interpreting is perceived as an unbiased practice, and interpreters are generally still expected to remain *neutral* by interpreting word for word and avoiding side conversations with the people for whom they interpret. On the other hand, advocacy is a practice that many scholars and practitioners accept in TPC and UX. And it is essential to Indigenous, Black, and Latinx scholars and practitioners in these fields who argue that engaging with underrepresented communities must involve advocacy and activism that directly benefit these communities (Itchuaqiyaq and Matheson 2021; Jones 2016; Rivera and Gonzales 2021). Thus, understanding Indigenous interpreters and translators as technical communicators in continuous engagement with Indigenous communities and examining their profession from multidisciplinary perspectives vindicates their advocacy work, foregrounding it as a major need.

Recognition and Respect

Nothing explains colonialism's effects and the power it still holds over public institutions in the Americas as well as Anibal Quijano's "Coloniality of Power, Eurocentrism, and Latin America" (2000). His work highlights the undeniable truth that Indigenous people continue to live under colonizing systems that persist in deeming Indigenous practices unreliable, abnormal, and rare. Coloniality has caused systemic issues that have kept Indigenous professionals on the margins of their own fields. Most Western public systems undervalue the work of Indigenous interpreters and translators, as pointed out by Natalia during her interview: "In the academic and government institutions, our work needs to be recognized as any other professional working as a translator or interpreter." Luis also pointed out that his daily challenge is "the recognition of the work of the interpreter" as he navigates court systems that undermine his profession only because he performs his job as an interpreter in his Nahua language. As I will cover in detail under the Issues section, Alejandro explained that part of the critical need to be valued as a professional includes "dignifying the job for interpreters" through better professionalization systems and dignified wages.

An important point to remember here is that Indigenous interpreters and translators are technical communicators who *translate* and *interpret* technical information (legal and medical, primarily) for nontechnical audiences. Yet technical and professional communication is associated almost exclusively with writing (and designing) in English. For this reason, social justice perspectives can help advocate for diversity by "interrogating how TPC can be complicit in reinforcing which perspectives and whose experiences are valued and legitimized" (Jones 2016, 2). Without a doubt, TPC has taken a tremendous leap forward in matters of social justice advocacy. Still, underrepresented languages, practices, and technical communicators must be part of these social justice conversations.

Updated Training

A major need that is part of the efforts to professionalize the field of Indigenous interpreting and translation is preparation in both theoretical and practical realms. TIS scholar Cristina Kleinert (2015, 2016) has centered her work on the need to refine and extend interpreting professionalization systems that address the specific needs of Indigenous interpreters in Mexico. She advocates for accredited professionalization systems as a critical factor for Indigenous interpreters to acquire agency.

Indigenous interpreters and translators who participated in the interviews identified training as a critical need. For example, Abril stated that her major challenge is "to continue to acquire, to get informed, and to take more updated training to continue helping people from [her] community." Gabriela explained that a need for her is training that can provide her with more cultural background: "To know about [Indigenous] cultures is what helps us with our work, and to know where we come from." Further, Mariana pointed out that her training needs to include training that can prepare her emotionally because she doesn't know "the situation that the job will bring day by day." Emotional training is critical because Indigenous interpreters must have the emotional maturity to withstand conflicting interpreting situations (Kleinert 2016). In the Issues section, I will address this topic in more detail as training is, evidently, one of the most significant needs for Indigenous interpreters and translators.

Use of Indigenous Practices during Interpreting Events

In chapter 4, I discussed in detail how using dialogue and having a pre-hearing time for a desahogo to become acquainted with the rhetorical situation Indigenous language mediators are about to interpret are important Indigenous practices that parallel practices promoted by scholars in different fields. Inghilleri (2012) in TIS proposes that we understand interpreting as a "living dialogue." My own work in rhetoric and composition has analyzed *desahogarse* as a climax cathartic point in testimonios that yields new possibilities to bring about social change (Rivera 2022b). Nevertheless, in practice, as explained throughout this study, Indigenous interpreters are still expected to perform a technical communication act without knowing much of the rhetorical situation from the point of view of the Indigenous person for whom they interpret, all while remaining stoic and *neutral*, as if neutrality were even possible.

Neutrality is a concept analogous to the common understanding of equality. For many, equality means providing the same treatment to all in order to be *fair*. And yet, equality—like neutrality—ignores that we can never be fair—nor neutral—if the rules of the game under which the *same treatment* is given privilege one culture. Walton, Moore, and Jones (2019) argue that in systems where privileges are not equal "and where oppression is hardwired into social structures, equality often cannot help but preserve the status quo, operating as a tool of oppression" (41). Equity, however, is a concept that embraces *adjusting* what is necessary to better address the needs of all as equally as possible, because addressing the needs of users involved in a T&I event taking place in a health institution in Mexico with an English speaker from the United States is different than addressing the needs of users involved in a T&I event taking place in an immigration detention center in the United States with an Indigenous speaker of K'iche' from Guatemala.

As in the case of the ideologies of equality, imbalances hide behind neutrality ideas that force Indigenous interpreters to remain stoic amid the clear non-horizontal communication interactions in which they participate. Indigenous interpreters are *rhetorical mediators* who privilege human rights and linguistic rights. They believe that injecting Indigenous practices can help balance the asymmetric interactions embedded in Indigenous interpreting events. Arguably, too, what Indigenous interpreters are pointing to is that no interpreter is truly unbiased. In the next chapter, I discuss equity in more detail and provide examples of ideation to show what some participants are doing to mitigate some of these issues.

ISSUES IDENTIFIED THROUGH TESTIMONIOS

In chapter 4, I discussed at length how testimonios are an Indigenous research method that better meets the needs of Indigenous Peoples and communities and can help when working with other underrepresented groups. Although scarcely used in technical fields, testimonios are a valuable method that can "help our TPC field by promoting both external and internal dialogues that reflect on individual experiences as they relate to our personal and professional communities and to our advocacy praxes" (Rivera and Gonzales 2021, 43–44). This section examines the pain points revealed through testimonio mapping. Because the complex issues mentioned here are analyzed from the perspectives of Indigenous professionals who work in various fields and in three different countries, this section centers on their voices. In chapter 6, I provide an overview of how TIS, TPC, and UX can help alleviate some of these issues. It is important to remember that two of the twelve interviewees are also among the nine professionals providing testimonios, hence the need to provide a second demographic table in this section.

As seen in table 5.2, the participants who shared their testimonios in the roundtable I moderated included six females, two males, and one nonbinary person; five participants were from Mexico, two from Peru, and two from the United States. Seven participants self-identified as Indigenous and two as non-Indigenous; seven worked in the legal field; four in the educational field; two in government initiatives; and one in the medical field. Eight different languages were represented in the roundtable, which included one speaker of Tzeltal de los Altos de Chiapas, one speaker of Zapoteco de la Sierra Sur, one speaker of Tarahumara, one speaker of Mixteco del Noroeste, one speaker of Chinanteco de San Pedro Yolox, nine speakers of Spanish, four speakers of English, and one speaker of Dutch. Some participants work in more than one field, and all of them speak more than one language. All participants worked with NGOs in different capacities. In chapter 4, I provided a full testimonio map of each one of the participants along with their biographical sketches.

The key issues, or pain points, that the participants identified include inconsistent local government policies, low and irregular wages, loose professionalization systems, discrimination, and lack of awareness about Indigenous matters. Eight of the nine participants in the roundtable identified lack of awareness and loose professionalization systems as critical issues, while five of the nine participants identified issues of low and irregular wages, local government policies, and discrimination (see figure 5.6).

Loose Professionalization Systems

By and large, participants indicated that loose professionalization systems that do not address the needs of Indigenous interpreters and

Table 5.2. Baseline characteristics of participants who shared testimonios

Baseline Characteristics	n	%
GENDER		
Female	6	67
Male	2	22
Nonbinary	1	11
INDIGENOUS BACKGROUND		
Indigenous individual	7	78
Non-Indigenous individual	2	22
COUNTRY		
Mexico	5	56
Peru	2	22
United States	2	22
FIELD[a]		
Educational	4	44
Government initiatives	2	22
Legal	7	78
Medical	1	11
LANGUAGES[b]		
Chinanteco de San Pedro Yolox	1	11
Dutch	1	11
English	4	44
Mixteco del Noroeste	1	11
Spanish	9	100
Tarahumara	1	11
Tzeltal de los Altos de Chiapas	1	11
Zapoteco de la Sierra Sur	1	11

a. *Some participants work in two or more fields.*
b. *All participants speak two or more languages.*

translators is one of the most prevalent issues in their field. Eight of the nine participants who shared testimonios mentioned this issue as a top priority. Having consistent, measurable skills in the profession not only impacts the quality of the services provided but also the status of the profession. Even with the complexities of working in a multidisciplinary, multilingual, multicultural, and multimodal field, Indigenous interpreters and translators believe that better standardizing systems to address their needs should exist across countries, modes, and professional fields. During our conversations, participants identified a series of professionalization-related issues that are specific to each of the fields in which they work. Therefore, in this section I synthesize the issue of professionalization by fields to provide a general survey of this problem in Mexico, Peru, and the United States.

**Issues
(9 Total Participants)**

Lack of Awareness	8
Loose Professionalization Systems	8
Discrimination	5
Low & Irregular Wages	5
Inconsistent Government Policies	5

■ Key Issues

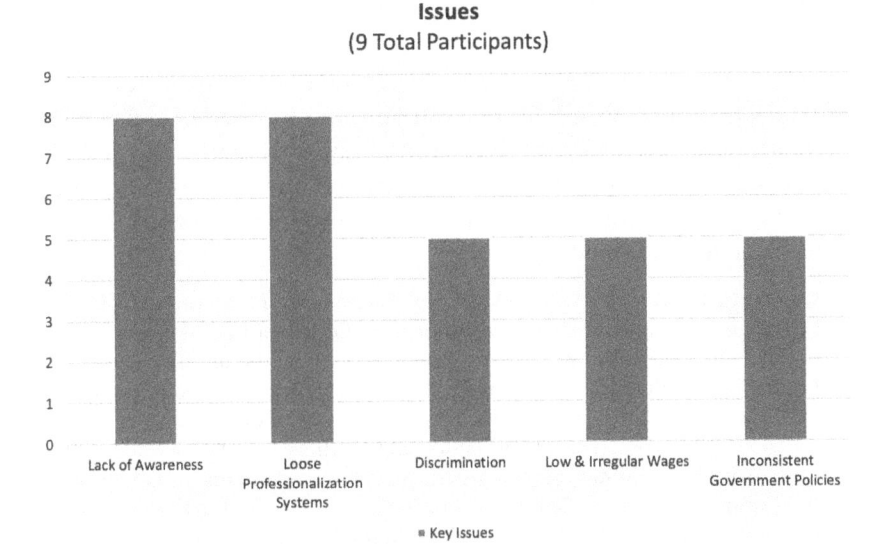

Figure 5.6. Issues identified by the participants in the roundtable

Legal Field

Court interpretation is an established profession in Peru, Mexico, and the United States, and thus it is the field in which most participants work. Extreme poverty, corrupt governments, crime, and the demand for illegal drugs have caused Indigenous diasporas in recent years. Consequently, international alliances between governments, interpreting agencies, and NGOs that help provide interpreting services for Indigenous people who find themselves in court systems are not uncommon. In spite of the frequent international collaborations, the certification requirements for court interpreters vary from country to country.

In Mexico, as Alejandro explained, there are three levels of court interpreter certifications with different requirements:

> In Mexico, there are certified, accredited, and practical interpreters. The majority of the interpreters are either accredited or practical. We only have one certified interpreter. These are our main issues, because in Oaxaca, as far as I know, there are only nine interpreters certified. Imagine! From the 176 variants spoken in Oaxaca, only nine interpreters are fully certified. Judges are required to look for certified interpreters first. Then, if there are none, they look for an accredited one, then for a practical interpreter. . . . We all believe that becoming a certified interpreter is important, but we have to see the other side of the issue. The only ones who certify at a national level are INALI [Instituto Nacional de Lenguas Indígenas] and CONOCER [Consejo Nacional de Normalización y Certificaión

de Competencias Laborales]. They certified interpreters in the years 2008, 2010, 2012, and 2014.

Certifying court interpreters who work with Indigenous languages comes with complex issues that in Mexico can only be addressed by Indigenous organizations, such as INALI, but having only one organization, clearly under-resourced, supporting the certification process of Indigenous court interpreters triggers other issues, like lack of professional follow-up, as Alejandro pointed out:

> We conducted a certification research in 2010, and we found out that after being certified, interpreters never went to a court hearing. Maybe it was because their linguistic variant wasn't required, or maybe another reason, but the issue is that no one followed up to ensure they continued to receive training and updates. I believe that it is not only important to certify Indigenous interpreters and translators but also there has to be a follow-up in the form of trainings, alignment courses, and assessments on behalf of the institution that provided the certification.... Who wouldn't want a certification? But not if they leave you alone in the journey without continued training and assessment. Practicums are needed, but you also have to know the interpreting techniques, like paraphrasing and emotional control.

As an experienced court interpreter, Alejandro also pointed out that because Indigenous interpreting events often occur under stressful circumstances, Indigenous language speakers are often inclined to *desahogarse* with the Indigenous interpreter (Rivera 2022a). However, Western interpretation protocols prohibit interpreters from engaging in any form of personal communication with anyone for whom they interpret. This leads to cultural issues among Indigenous language speakers and their interpreters and imposes excessive stress on both during a translating event. Alejandro wishes protocols would allow Indigenous interpreters to interact with Indigenous defendants before the formal interpretation begins to better understand the context of the situation and to better manage the desahogo, because it inevitably happens during the interpretation. This aspect of Indigenous interpreting events is not addressed in the certification training offered by agencies (governmental, nongovernmental, and private) as they continue to endorse neutrality.

In Peru, Indigenous interpreting and translation certification programs are also provided by the government, as explained by Lourdes:

> Indeed, the State is the one who offers the accreditation.... The certification operates on two levels: the accreditation of taking a course and the administrative process that comes with completing a series of practicums in a public institution. And finally, the registry.... Any public institution, the Health Ministry or the Justice Ministry, can access this registry and see in the list the language for which an interpretation or translation is needed.

Lourdes pointed out that more than a certification in translation and interpretation of Indigenous languages is needed to conduct effective legal and medical services. Follow-up training in the professional field is necessary.

> There are intensive courses given every semester, and after a certain number of hours, that Indigenous translator and interpreter is accredited. Then, there's a national registry, but that isn't enough because Indigenous interpreters have to be professionalized. In other words, Indigenous interpreters haven't necessarily completed a professional degree; they aren't necessarily healthcare experts, they aren't necessarily court process servers, and they aren't necessarily linguists, although there are some. This doesn't mean that there are none. Most likely they have been an Indigenous leader within their community, appointed by the community, and sent to Lima, the central authority, to take this course. . . . About the training given by the Ministry of Culture, some people say that it is intensive, but there are cases, for example, where someone hasn't received a postsecondary education that allows this person to understand a Western system and judicial administration enough to comprehend what is an advanced process, commutation of sentence, or simply matters of intercultural healthcare, like what it means to be given anesthesia. For some Indigenous women, to be under anesthesia is to die.

The United States also has a court interpreter certification system provided by the government. However, there has yet to be a certification available for Indigenous languages of Latin America despite the demand in recent decades. To become a court interpreter of Indigenous languages, interpreters must demonstrate language skills in three languages—English, Spanish, and their Indigenous language, as explained by Julia:

> In the United States, there's no certification for Indigenous languages of Mexico. There are legal and healthcare certifications at the national level, but only for the three predominant languages, which are Spanish, Chinese, and another language. . . . In the legal field, there are three categories for interpreters: certified, professionally qualified, and skilled.

Julia also emphasized another barrier that Indigenous interpreters in the United States face when working for the courts. To become a certified court interpreter in the United States, they must have a minimum of high school–level education in English.

> At the legal and national certification level, there is also a prerequisite, which is to have at least taken high school. This is a criterion that many of our interpreters do not meet. And in the US, this has to be met in English. You can take your high school courses here in Mexico, but you also have to take the GED, a high school equivalency test for adults, which is administered in English, so that you can meet the education requirements to practice in the United States.

Medical Field

Although there are certification systems for Indigenous interpreters who work in the medical field in Mexico and Peru—the same as court interpreters—it is not common for healthcare providers to hire the professional services of Indigenous interpreters in these countries. In the medical field, Indigenous interpretation occurs, but at an informal, mostly unpaid level. Although there is a more robust system of healthcare interpreters in the United States, interpreters of Indigenous languages working in this industry must demonstrate proficiency not only in their Indigenous language but also in both English and Spanish, as in the legal field. The certification processes that healthcare interpreters must undergo in the United States are managed by national nonprofit agencies, as explained by Julia.

> They offer what's called 'CoreCHI' [Certification for Healthcare Interpreters]. You can still get certified [in other languages], but you have to know English and Spanish. You can be an Indigenous person who speaks an Indigenous language and still get a national certification in the healthcare industry. Still, you are required to know English and Spanish. What's happening in our communities is that most speakers of Indigenous languages only speak Spanish. [The healthcare accreditation consists of] attending a course with a certain number of hours, which covers standards, ethics, and protocols based on the guidelines of the agency CCHI [Certification Commission for Healthcare Interpreters]. If anyone wants to become certified in that field, then they have to pay 500.00 dollars. I believe that there's only a handful of interpreters in the State of California who have this national certification in the healthcare industry.

Julia also highlighted the issue of following up and having evaluation systems for Indigenous interpreters who work with Indigenous languages that are rare in the public systems of the United States.

> We want to establish networks for evaluation since sometimes there aren't even three people who speak the same language, and we don't want to be the only ones who can certify ourselves. We want to learn more about this process for all of us who speak Mixteco in their own language variant because nobody else will be able [to certify]. Other people do understand the language, but sometimes they don't understand it well enough.

Gaining experience through practicums is another issue that Julia, and Indigenous interpreters working in other fields, noted: "There are different perspectives on what a certification really means, because sometimes those who are certified, despite them having a certificate, do not understand things the same as those who have done a practicum."

It should be noted that although the United States has systems in place to provide interpreting services for healthcare users, it is still very common to rely on a patient's family member(s)—sometimes children—to provide an interpretation or to *get by* with words and phrases that doctors and nurses know in commonly used languages, like Spanish.

Educational Field

As in the case of medical interpreters, professional interpreters in the educational field are rare in Latin American countries like Peru and Mexico. Interpretation in Indigenous languages in this field occurs often, but, again, mostly informally and without pay. It is often done by bilingual teachers or children (in chapter 3, I discuss the relationship between Indigenous interpreters and child language brokering). In schools in the United States, interpretation between commonly used languages, like Spanish, is typically done by bilingual educators. However, because educators of Indigenous languages are not common in schools, districts have come to rely on nonprofit organizations working with Indigenous migrant communities to find and train Indigenous interpreters.

And yet, school districts in the United States find it challenging to locate Indigenous interpreters who can also communicate in both Spanish and English, as explained by Valeria:

> Because I have to work with communities that are neither fluent in English nor Spanish, I have to recruit any interpreter available, whether they are accredited or certified or have experience. Either way, I am going to do what I can to communicate with Indigenous families.... Once Indigenous parents find out that we have an interpreter who can help, there's more demand, and right now we don't have enough people who can interpret, and we don't have enough training available in our city either.

Valeria also pointed out that although it is difficult for school districts to find Indigenous interpreters, it is important to refrain from having children interpret for their parents: "We recruit interpreters over 18 years old, and we try not to recruit students because we don't put students younger than 18 years old in traumatic situations. It is very difficult for students to try to interpret for mom and dad a painful situation in which they were involved."

Government Initiatives

The absence of adequate professionalizing systems that address the needs of Indigenous interpreters also causes issues in government initiatives, even when involving a university specialized in the study and the practice of translation and interpretation. Such was the case of a project where Magdalena was involved. In this project, the Amazonian Indigenous community of Achuar del Pastaza in Peru asked the government to translate civil registry certificates into their Indigenous language. Although the government agreed to provide bilingual birth, death, and marriage certificates, it did not have enough resources to do so. Therefore, the government asked the Translation and Interpretation program at the Universidad Peruana de Ciencias Aplicadas (UPC) (University of Peru of Applied Sciences) for help.

As Magdalena explained, the university did not have Achuar translators, so the project organizers sought out the help of "two Achuar educators, who were not interpreters or translators. They were bilingual educators but took on the role of professional translators." I provide a more detailed account of this project and Magdalena's testimonio in chapter 6. While the issue of not having trained translators in the Achuar language was apparent, other critical issues emerged during this project, as Magdalena pointed out: "Because UPC was used to working with foreign languages, they did not want to go to the Achuar community. . . . For the first time, two Achuar people came to UPC. . . . Imagine! The academia, which only teaches Chinese, French, German, and Italian, suddenly brings Indigenous educators with a different language, a different worldview, a different cosmovision."

It is assumed that academia, as an institution designed to assist in the professionalization of different fields, would help with the professionalization of Indigenous interpreters and translators. Unfortunately, as Magdalena highlighted, there is a clear lack of attention to the translation and interpretation of Indigenous languages on behalf of educational institutions as they noticeably privilege Western languages and Western practices.

In the same way that most translation and interpretation academic programs focus on languages that produce professionals who could work in more profitable industries, like tourism and business, most technical communication and user experience academic programs prioritize fields that engage with new technologies. Therefore, the professionalization of Indigenous interpreters and translators working in the legal, medical, and educational industries has become primarily an endeavor of Indigenous organizations.

Low and Irregular Wages

Five of the nine participants who shared their testimonios identified issues related to wages in their profession. As in the case of any other profession, wages in the interpretation and translation field vary country by country. Nonetheless, there are also clear, abnormal differences among industries and, in the case of court interpreters, between federal courts and local courts. As explained before, Indigenous interpretation and translation in the medical and educational fields are not often paid services in Latin America, hence why most Indigenous interpreters and translators work in the legal field.

One of the issues affecting Indigenous court interpreters is that many of them come from rural communities and thus must pay travel expenses to attend court hearings, as Alejandro explained.

> The federal court in Oaxaca pays 1,800 Mexican pesos per hearing for up to 3 hours, and sometimes they pay for travel expenses. . . . The common courts pay only 400 Mexican pesos per hearing and that includes travel expenses. . . . Imagine the expenses of an interpreter who comes from Coixtlahuaca to Salina Cruz. From Coixtlahuaca to Oaxaca City you have to spend about 200 pesos. From there, you have to take a car to Salina Cruz, which is another 120 pesos. That's a total of 320 pesos, but that is just one way, and that is only transportation, and it doesn't include food and lodging, then it is not enough. . . . And if the interpreter who made the commitment to attend doesn't show up, then this interpreter is fined.

Indigenous court interpreters grapple with irregular and low wages in other states of Mexico too, as explained by Antonia. "There are many issues, like irregular wages. They aren't the same everywhere. In Chihuahua, they pay 200 Mexican pesos per hearing, and someone else said they pay 400 in Oaxaca, and then 1,800 in the federal court. Why does it vary like this?"

Lourdes pointed out that this is also an issue in Peru. Although Lima is the home of some Indigenous communities, most Indigenous court interpreters live in rural communities far away from the capital: "A problem transpires though, what happens if the interpreter comes from a Peruvian Amazonian Achuar community and has to take first a boat of various hours of river journey to later arrive in Lima, and then take a plane [to the final destination]? Who pays for the travel expenses? Who pays for the transportation? The reimbursement isn't clear. A plane ticket costs 150 dollars minimum."

Additionally, the perception is that an interpreter and/or translator of Indigenous languages receives lower wages than their counterparts who work with Western European languages. As Claudia argued, "The wages for Indigenous interpreters aren't defined. Are we supposed to receive less than an interpreter of a foreign language? Because interpreters of foreign languages get paid very well."

Two of the nine participants who shared their testimonios also commented that payment for the services of Indigenous court interpreters is often delayed for months or even years. Claudia explained that in Mexico City sometimes "the problem is that you go and translate or interpret, but they don't pay you until three or four months down the road. In fact, they owe us since 2015." In Oaxaca, Alejandro has experienced similar issues: "Sometimes the courts don't want to pay Indigenous interpreters at the end of the hearing, nor do they want to give interpreters the paper that proves that they attended the hearing. . . . Then, they owe money to many interpreters. They owe to interpreters from [our organization] and others who work independently, who have told me that the courts owe them money since 2017 or 2018."

While the payment delay seems also to be a problem for court interpreters in the United States, this country adheres to more standardized payment policies for interpreters working in both the legal and the medical fields, but, as explained before, the problem for Indigenous interpreters is to pass the difficult certification exams in both English and Spanish. In fact, because of these strict requirements, there is a shortage of interpreters who work with Indigenous languages from Latin America in the United States. For this reason, it is not uncommon for the US courts to contact Indigenous organizations in Mexico or other Latin American countries to help find an Indigenous interpreter who can work with a specific language variant not available in the United States. When this happens, the Indigenous interpreter has to do the work through a video conferencing application or via phone, depending on the technology available in the place where the Indigenous interpreter is located.

Providing Indigenous interpreters for families of Indigenous students attending US schools is not a common practice because it wasn't until the massive Indigenous diaspora of recent decades that the need became visible. Yet, districts in places where Indigenous migrant communities concentrate, like California, are increasingly trying to accommodate the needs of Indigenous families by hiring in-house Indigenous interpreters within some school districts. Valeria pointed out that although school districts might have the funds to hire an Indigenous interpreter, interpreters of Indigenous languages are not readily available, and thus organizations that work with Indigenous migrant communities, and sometimes the same districts, embark on a long process of finding a person within an Indigenous community who can help as an interpreter informally and later on be formally trained as one:

> I work with interpreters who want to help others from the heart and who are willing to work voluntarily, those who say, "I'll go whether they pay me or not." Then, I look for training and ways to ensure that the district

pays for the professional development of the interpreter. Once this interpreter has received the appropriate training, the district hires this person as a full-time interpreter. . . . The school district has the funds, and the key person in the school district and I work together to ensure that the interpreter gets paid.

Evidently, the issues of low wages for Indigenous interpretation services in Latin American courts are connected to the low government budgets in developing countries, and payment issues might be linked to corruption in government agencies. Yet, there is another cultural factor that, according to Mónica Morales-Good (2020), comes into play in conversations about low wages: the *tequio*. As discussed throughout this study, Indigenous interpreters and translators, like most Indigenous individuals, are committed to helping their Indigenous communities. This Indigenous communal work is known in Mexico as tequio, a concept with which most public officials working in places with sizeable Indigenous populations are familiar. Morales-Good points out that sometimes public officials pressure Indigenous interpreters into providing their services for free as a form of tequio. "Not all interpreters are able to donate their time and skills, and they shouldn't be expected to do so," Morales-Good argues (163). To offset this problem, Indigenous organizations, such as CEPIADET, have initiated campaigns of *concientización* that aim to create awareness among public officials about issues like this.

Inconsistent Local Government Policies

As previously noted, most of the Indigenous interpreters and translators in this study understand agency as advocacy. Many engage in activism through Indigenous-led organizations and thus feel affected by fluctuating local government policies. Five of the nine participants in the roundtable believe that constant changes in the local government affect the work of Indigenous interpreters and translators both as practitioners and advocates. In Mexico, for instance, state governments change every six years, and each government brings a new cabinet and new policies. As a result, initiatives by one government are not always carried out by another, as Alejandro pointed out. He praises the way that the Defender's Office in the state of Oaxaca is now actively hiring more Indigenous defenders, but he worries that new incoming governments might not sustain this enthusiasm toward addressing Indigenous matters. "It is important what the Defender's Office is doing" because "not all incoming governments have that perspective, and sometimes they abandon Indigenous topics," Alejandro commented.

When there is a shift of power, Indigenous interpreters and translators are affected even if laws remain unchanged. As Antonia explained, "sometimes this happens because they are temporary governments. Those who come in, arrive without knowing how to continue the work or the commitment that others made." Although courts are supposed to provide interpreters for victims and defendants who do not speak Spanish, Antonia argued that "those who work for the justice departments don't care. They look for people [to interpret] on the streets, even if they aren't certified. Sometimes they even look inside the CERESO prison."[1] Similarly, Claudia noted that the inequities are also visible in the government spending: "I don't know if you all heard the speech by the Judge of Morelos. There I saw the difference between the foreign and the Indigenous, and the discrimination. He said that an economic ruling was asked, and the surveyor was paid 24,000 Mexican pesos. Then, he said that an anthropological ruling was asked, and that surveyor was paid 12,000 Mexican pesos. What is happening?" Claudia believes that the Mexican government budgets more resources to initiatives that, from the lens of the government, promote progress rather than assigning more resources to Indigenous matters, which are regularly considered anthropological matters.

In Peru, the government has made provisions to exercise Convention No. 169, which requires that the government and private investors consult with Native Peoples before embarking on initiatives near their communities. However, as explained by Lourdes, the state establishes these provisions "from the needs of the State, and not of the Native Peoples." This provokes other critical issues for Indigenous interpreters and translators. Lourdes pointed out that "in cases of prior consultation, for example, Indigenous interpreters can be seen as people who help the process of the State and not as someone who is really conveying the needs of the Indigenous community." The Peruvian government does not even acknowledge them as Indigenous interpreters. "For the State," said Lourdes, "they are 'interpreters in native languages.' However, some interpreters want to be recognized as *Indigenous interpreters*."

Additionally, Indigenous interpreters and translators believe that local governments should do more in matters of training and wages, particularly in the case of court interpreters. Governments have systems in place for other professionals, and Indigenous interpreters wish governments had more involvement in providing solutions to their critical issues.

Discrimination

Language is not what most people consider when thinking of discrimination. While there have been efforts to raise awareness about gender, religion, age, and disability discrimination, we tend to primarily link this unethical practice to the unequal treatment of a person based on race and/or ethnicity, overlooking the strong connection between language and race or ethnicity. As most participants pointed out, discrimination has caused the marginalization and decline of Indigenous languages. Five of the nine participants believe that discrimination is a palpable issue among Indigenous translators and interpreters.

Take, for instance, Carlos. He shared that in his community, "most of the town [elders], about 90 percent, speak an Indigenous language, Zapoteco. Adults suffer discrimination when they move to the city or work here and there." It is because of discrimination that Carlos tried not to speak his Zapoteco language in school at first: "When I was at school, I was embarrassed to speak it." Carlos added that "nowadays in [his] community the majority are losing [the Zapoteco language], more so the kids, because their parents are teaching them Spanish, because of the same reason, parents don't want their kids to suffer embarrassments or discrimination by other kids whose first language is Spanish when they arrived at the city." As an interpreter, Carlos helps incarcerated Indigenous people better understand the legal system. As an intercultural promoter, he teaches Zapoteco to the children in his community: "These are kids who are learning Spanish and are letting go of their Indigenous language." He became an advocate "because the majority of the men in [his] town leave to the East or West Coast of the United States." Carlos helps his community and his culture through language.

Discrimination against Indigenous people happens in different spaces, as seen in the testimonio presented by Lourdes. She talked about a court case in Peru that ruled in favor of an Indigenous woman speaker of Quechua. The court made the Indigenous woman sign a document that only allowed her to prepare food in public spaces at certain hours of the day in a case that initially addressed this woman's right to work. The judge found out that the woman had signed with her fingerprint, which meant that she did not speak Spanish and signed a paper in Spanish without understanding it, and thus ruled it as a language discrimination case. Lourdes explained that "the court said, 'this is not a case of the right to work, it is a case of the right to not being discriminated against for using a particular language.' ... And stated that this was a discrimination case because there was no equity."

Sadly, discrimination against Indigenous people is not a new problem. It has existed for a long time outside and inside Indigenous communities; as Antonia asserted, "we have heard about human rights and about discrimination for a long time, and we are still talking about the same issues with no solutions. Sometimes there's even discrimination among ourselves. What should we expect from those who aren't of our race?" While there have been efforts to combat this issue, discrimination continues to be one of the most critical problems affecting Indigenous people and, therefore, Indigenous interpreters and translators.

Although the participants who spoke about discrimination during the roundtable did it mainly in terms of language, it is clear that the issues of low and irregular wages, local government policies, and loose professionalization systems mentioned before are also entwined with more systematic issues of discrimination.

Lack of Awareness

Lack of awareness about Indigenous matters is another issue that nearly all participants indicated. Eight of the nine participants pointed out that there needs to be more awareness about Indigenous linguistic rights and about the problems faced by Indigenous people overall. Most participants explained that the lack of awareness about Indigenous matters among public officials and public workers causes many severe issues for Indigenous people. Therefore, they believe that the role of Indigenous interpreters and translators goes beyond language brokerage. Their role includes creating awareness about Indigenous matters.

The project in which Magdalena participated comes to mind, where the government of Peru worked with the Department of Translation and Interpreting Studies at the Universidad Peruana de Ciencias Aplicadas to translate birth, death, and marriage certificates to Achuar. Even though this project provides a great example of intercultural and interinstitutional work, Magdalena questioned why it is taking governments over five hundred years to translate certificates that validate a human being's identity into a language that can be understood by the person who holds it. In her testimonio, Magdalena explained how the Amazonian town of Achuar del Pastaza asked the government to provide them with bilingual "vital event" documents, such as birth, death, and marriage certificates: "These certificates are made by the Registro Nacional de Identificación y Estado Civil [RENIEC], you must have an equivalent here, which is the institution in charge of giving your DNI [Documentación Nacional de Identidad (National ID)]. If someone dies, there is a certificate. And also, if someone is born, so that this person can acquire a DNI." Magdalena explained that involving Indigenous translators helped her institution recognize the need for creating alliances between the academia and Indigenous communities to help and learn from one another. I discuss this project in detail in the next chapter.

Furthermore, as Rosa implied in her testimonio, the lack of awareness about Indigenous linguistic rights affects Indigenous defendants' perception of everyone involved in a court case, including interpreters. When she first became involved as an intercultural promoter and interpreter, she realized that most Indigenous defendants go through negative court experiences and thus saw her with distrust: "During the visits to the reintegration centers, we realized that people didn't come to us, not because they didn't want to but because there was no communication bridge because, well, if you can't even defend yourself in Spanish little less speaking an [Indigenous] language." She explained that it is important to create awareness about Indigenous linguistic rights among public officials because in Oaxaca, for example, "the Public Defender's Office legally represents 98 percent of Indigenous people who are deprived of their freedom in different reintegration centers in the state," and there are only a few public defenders who speak Indigenous languages, hence the importance of hiring Indigenous interpreters in places with a large Indigenous population.

Although Indigenous language rights are protected by several national and international laws, in practice, not all are known by Indigenous people, and not all are respected by local governments, as Alejandro pointed out:

> There are many laws in Mexico that address Indigenous language rights and the right to have an interpreter in legal procedures, like the Federal Law of Linguistic Rights of Indigenous Peoples, articles 9 and 10, the second article of the Mexican Constitution, the National Code of Criminal Procedures, and the agreement 169 of the OIT [Organización Internacional del Trabajo (International Labor Organization)], but sometimes the reality is another thing.

Lourdes also explained that in the last decades, laws that protect Indigenous rights have become more visible in Peru: "The law establishes the right, and I quote, 'to enjoy and to have access to translation means . . . that guarantee the enactment of rights in all sectors.' . . . But this is only one step." She also said that international laws, specifically the Convention No. 169 by the International Labor Organization [ILO], were the reason that triggered the Peruvian government to create a certification program for Indigenous interpreters and translators:

> In Peru, the certification and accreditation of interpreters of Indigenous languages was created following the Convention No. 169 [of 1989] by the

International Labor Organization, and it draws on the Native Peoples' fundamental right to express in their own language when discussing matters with the judicial authorities of the State. Convention No. 169 requires that Native Peoples be consulted in different matters. To make provision for this, a law of prior consultation, in their own language, was established.

Like Alejandro, Lourdes emphasized that despite having laws such as this, governments still perceive the law "from the needs of the State, and not of the Native Peoples."

Claudia also pointed out that the issues affecting Indigenous people are sometimes more accentuated within Indigenous migrant communities. Claudia belongs to a Tzeltal community from Chiapas but has lived in Mexico City for over twenty years. As an Indigenous migrant, she sees the indifference of the local government toward her Tzeltal migrant community in the city. During her testimonio, she explained how other migrant Indigenous communities in Mexico City have faced the same issues and informed us that after the Triqui community challenged the indifference of the local government toward their problems with violence and discrimination, the local government paid more attention to other Indigenous migrant communities in the city:

> There is a constant struggle. But since the Triqui Indigenous movement in Mexico City took place, the public policies in Mexico City have paid more attention to the Indigenous migrant population . . . because when there are Indigenous migrants, the local government washes its hands because they think that's not their problem.

Like Claudia, Valeria has also seen similar indifferent attitudes toward Indigenous migrant communities in the United States. Because some of these communities seem to be invisible within public systems, she advocates for Indigenous language rights in the school districts in the community where she lives: "The most important thing for us, for the interpreter, for the district liaison, and for me, is to ensure that Indigenous immigrants are heard in the school districts." In chapter 6, I discuss how a lack of awareness about Indigenous matters can, and often does, provoke serious discrimination issues.

CIVIC ENGAGEMENT ACTIVITIES IDENTIFIED THROUGH TESTIMONIOS

Testimonio mapping also revealed the important civic engagement activities in which participants take part to counteract the issues in their profession, as shown in figure 5.7. Of the nine participants, six revealed

Civic Engagement Activities
(9 Total Participants)

Mentoring Projects for Interpreters	6
Indigenous Rights Activism	3
Language Revitalization Projects	3
Intercultural Translation Projects	1

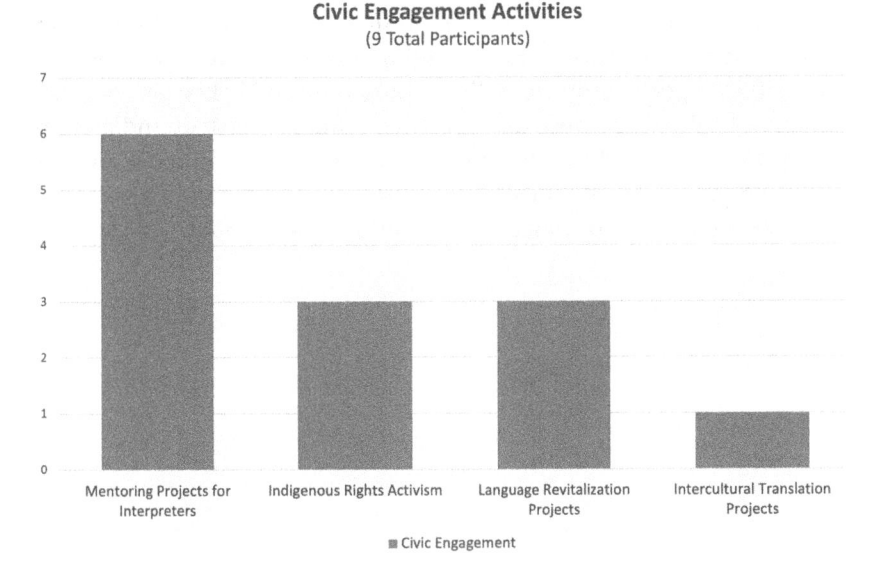

Figure 5.7. Civic engagement activities in which participants take part

that they take part in programs that mentor Indigenous interpreters and translators, preparing them not only with linguistic and technical strategies but also with information about how to navigate Western public systems to help their communities; three participants engage in Indigenous rights activism; three participants engage in Indigenous language revitalization projects; and one participant is involved in intercultural translation projects that bridge collaborations between universities, government institutions, and Indigenous communities.

Each participant engaged in activities that reflected their context. For example, younger professionals engaged in language revitalization projects that capitalize on new technologies, like creating videos to help Indigenous children become more fluent in their mother tongue. Conversely, seasoned professionals with more years of experience engaged in projects aimed at mentoring Indigenous interpreters and translators. Professionals with more access to resources (and to the English language as a resource), like academics and individuals working for nonprofits in the United States, engaged in Indigenous rights activism. The one participant from the roundtable who engaged in written translation projects also came from an academic background.

In the next chapter, I discuss three important projects presented by the participants in this study: an intercultural and interdisciplinary translation project involving Achuar translators in Peru, a project led by young interpreters and intercultural promoters from diverse Indigenous communities in Oaxaca, Mexico, and a project that aims at teaching young P'urhepecha girls the skills of translation in Michoacán, Mexico. Chapter 6 also examines the contextual interactions of Indigenous interpreting events through a user interface model and analyzes how placing equity rather than usability at the core of UX can help alleviate the systemic issues with which Indigenous people grapple.

6
IDEATING AND RE-DESIGNING

As discussed in chapter 5, the more palpable issues of professionalization, low wages, and unstable government policies stem from the systemic issue of discrimination (see figure 6.1). And whereas discrimination can grow from ill will, most prejudices are caused by a lack of understanding or by false impressions of something or someone. The Indigenous participants I worked with during the unconference were well aware of this. Therefore, they believed that a key strategy to address these issues is to raise awareness about Indigenous cultures, languages, and laws that protect them, their practices, and their languages. As seen in figure 6.1, the solutions proposed to create awareness include advocating through technology, building alliances with academia, and supporting Indigenous language revitalization projects.

This chapter covers three major projects shared by participants through their testimonios, which involved the participation of Indigenous interpreters and translators in projects in collaboration with public institutions. I also discuss some ideas the participants proposed as possible solutions. But first I would like to cover a detailed analysis of the importance of placing equity in UX research.

EQUITY IN INTERPRETING AS A TECHNICAL COMMUNICATION PRACTICE

Understanding Indigenous translators and interpreters as technical and professional communicators helps us see their need for advocacy as they argue for acknowledging communication practices that address the whole human experience (Jones 2016). And after seeing the systemic issues that affect Indigenous interpreters and translators, the answer to my third research question became apparent. Why is it important to place equity rather than usability at the core of UX research? Because the traditional model designed by Peter Morville (see figure 6.3), albeit of great significance, does not address the needs of all users. Marginalized groups, as seen in this research project, often face systemic issues

https://doi.org/10.7330/9781646425310.c006

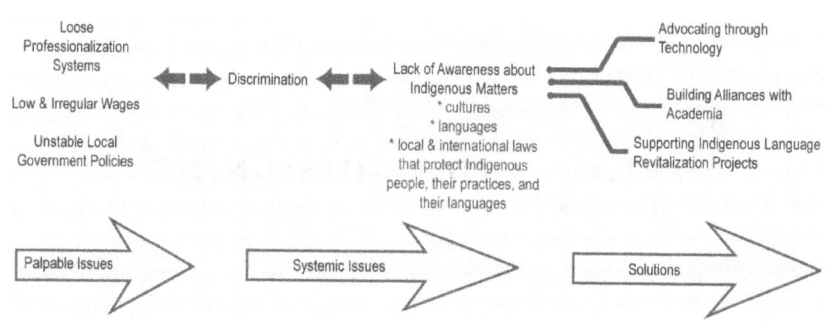

Figure 6.1. Issues and solutions

Palpable Issues:
- Loose Professionalization Systems
- Low & Irregular Wages
- Unstable Local Government Policies
- Discrimination

Systemic Issues:
- Lack of Awareness about Indigenous Matters
  * cultures
  * languages
  * local & international laws that protect Indigenous people, their practices, and their languages

Solutions:
- Advocating through Technology
- Building Alliances with Academia
- Supporting Indigenous Language Revitalization Projects

associated with relationships of power, which can only be addressed by designing more equitable products, contents, and processes. During our collaborative event, it became clear that equity, as the factor that addresses relationships of power, is the single most important factor for Indigenous users. Most importantly, this research shows that many times to foster equity we have to differentiate approaches. As I explained in chapter 5, equity should not mean addressing the needs of everyone in equal ways. It means *adjusting* what is necessary to address the needs of everyone as equally as possible. And thus, as shown throughout this study, we cannot examine Indigenous translation and interpretation in the same way we examine these events using European languages.

To fully comprehend what is at stake in an Indigenous interpreting event, I created a visual model of the *user interface* (UI) of Indigenous interpreting events (see figure 6.2). User experience researchers use this tool to visualize the space where users interact with technology. I use the UI to visualize the relationships involved in Indigenous interpreting events as a means to analyze the interactions between linguistic, technical (specialized terminology), cultural, institutional or traditional, and sociopolitical contexts.

Communication models usually depict horizontal sender-receiver interactions. However, as shown in figure 6.2, this UI has a significant imbalance because the interactions happen in a space guarded by a Western cosmovision. As Indigenous interpreters very well know, interactions between Indigenous and non-Indigenous users are not horizontal, especially in highly hierarchical spaces such as courts. This uneven relationship is what Indigenous interpreter and activist Edith Matías Juan (2021) calls "institutional verticality," which produces not a horizontal dialogue but an unfair vertical discourse. It should not come as a surprise that centuries of bestowing a higher value on Western languages, cultures, and institutions have provoked asymmetric discourses today.

What might not be apparent is how the space in which we interact and the pervasive use of technical and scientific jargon in our institutions also accentuate these inequities. Indigenous interpreter Guadalupe Pérez Holguin argues that technical language makes court interpreters feel belittled when communicating with court officials:

> The language of the judge is very complicated. We interpreters must be courageous and not be afraid of anyone. Not of the judge, the prosecutor, or anyone. We should instead say, "Your Honor, could you please reformulate the question? Could you make it simpler? We can't understand you." ... One feels small next to them. But why do we feel that way? Because there's no [horizontal] communication. (Blanco 2018)

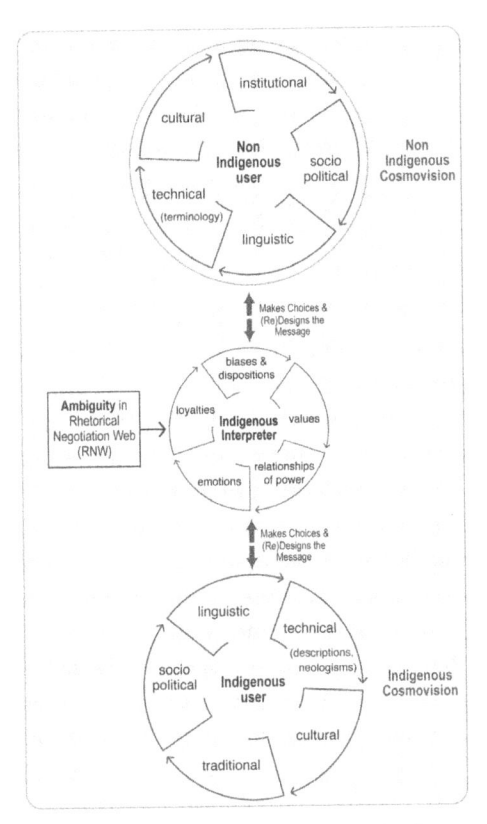

Figure 6.2. Vertical user interface (UI) of Indigenous interpreting events

Many of these imbalances hide behind ideas of *neutrality* that expect Indigenous interpreters to remain stoic amid the inequities they witness and behind a *code of ethics* that only protects institutions rather than relying on professional ethics that privilege human and linguistic rights. Notwithstanding, in practice Indigenous interpreters are *rhetorical mediators* who navigate a *rhetorical negotiation web* (RNW) to negotiate truths, or meaning, by making rhetorical choices in order to re-design a message, all while mediating the values, emotions, loyalties, relationships of power, biases, and dispositions of the users and of themselves. And because Indigenous interpreters mediate messages and negotiate between users with very different values, emotions, loyalties, relationships of power, biases, and dispositions, interpreting events inherently yield *ambiguity*. For example, in chapter 3 Mariana explained that she prepares emotionally because she does not know "the situation that the job will bring day by day" (table

3.2), showing that emotions are always involved in an interpreting event. Victoria pointed out how the job of an Indigenous interpreter and translator is "a double-edged delicate matter" because she mediates information between the holders of power, the government, and her community members and, therefore, her job can help but also hurt her community (chapter 3, table 3.4). Natalia also explained that her reason for being in the field "has always been to watch over [her] people in all its cultural aspects" (chapter 3, table 3.6), proving that there are always loyalties and biases involved. Claudia sees this profession as a "calling" because of the communication needs of her community (chapter 3, table 3.7), demonstrating that power imbalances, and thus injustices, in communication happen all the time, whether Western public institutions recognize them or not. Western cosmovision, nonetheless, continues to subscribe to the Aristotelian notion of ambiguity as a "vice of language" (Golitsis 2021). This Western view of ambiguity is closely associated with the outdated machine-like understanding of interpretation that sees language as if it existed in a vacuum, voided from its contexts and without a human to perform it. Indigenous interpreters, in one way or another, understand that communication, specifically technical communication, never happens without the presence of asymmetry and ambiguity, hence the importance of fostering equity to adjust it.

Indigenous interpreter and activist Abigail Castellanos García (2021) believes that the solution to challenging institutional verticality has to come from within public institutions by generating internal institutional policies that not only embrace Indigenous practices and languages but also encourage their use. Similarly, Matías Juan (2021) argues that laws that promote Indigenous linguistic justice are not enough to make significant changes. For her and for many other Indigenous activists, a formal linguistic plan that regulates the teaching of Indigenous languages and the college education of Indigenous interpreters and translators is the key. While there is an evident need for TIS to expand the field to include Indigenous perspectives in their academic offerings, TPC too needs to be accountable for providing more resources for underrepresented groups who struggle to navigate Western spaces—foreign spaces for many—with highly specialized terminology that continues to maintain oppressive institutional vertical systems.

EQUITY IN COMMUNITY-BASED UX RESEARCH

Since its inception, UX research has aimed at understanding users' needs and motivations. It subscribes to the idea that to produce valuable

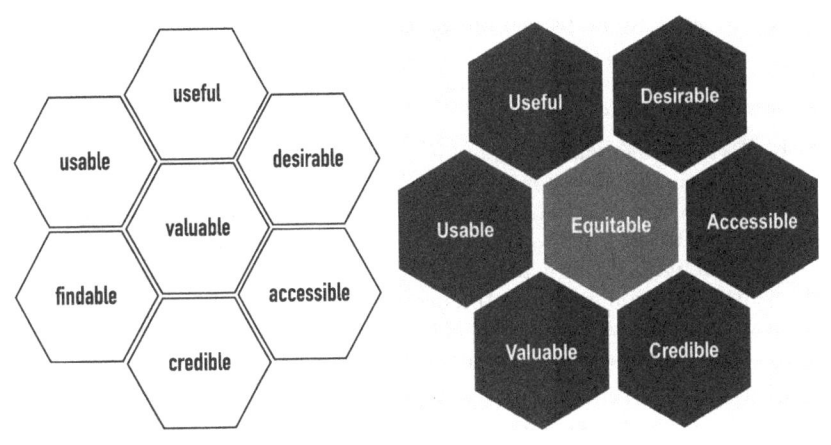

Figure 6.3. Peter Morville's (2014) user experience honeycomb

Figure 6.4. User experience honeycomb with equity factor at its core

products, processes, and content, these must be useful, usable, desirable, findable, accessible, and credible, as shown in Peter Morville's (2014) model in figure 6.3 (also known as "The Honeycomb"). This model, however, fails to acknowledge the importance of equity as a factor that addresses the power relationships between users. This study proposes an adaptation of Morville's Honeycomb by removing "findable" (because this factor did not apply to the physical space in which the event took place) and adding "equity" as the core factor in this UX project, as illustrated in figure 6.4.

The Indigenous organization leading the production of the unconference, the Centro Profesional Indígena de Asesoría, Defensa y Traducción (CEPIADET), had a clear vision of the factors that needed to be considered throughout the co-construction of the event, for they had done something similar at a lower scale previously and wanted to provide a more meaningful experience. Although the factors addressed in UX are typically applied to online spaces, they align with what we designed in the physical space of the event. As a team, we designed a conference experience anchored in the following interconnecting factors.

a. Useful. In UX, practitioners refer to this factor as the one that aims at designing content, products, and processes that fulfill a need. In the case of translation and interpretation systems (professionalization programs and public sector regulations) in the Americas, they continue to be largely based on systems imposed by Europeans and thus rarely take into account the needs of Indigenous users (interpreters, translators, and users of public sector services). Therefore, the event provided the space for participants to generate strategies that can help fulfill the needs of Indigenous users both as translators and interpreters and as users of public sector institutions in need of translation and/or interpreting services.

b. Usable. The usability of products, services, and content is still the most important factor of UX research. Its main concern continues to be easy-to-use digital interfaces. Yet, more than solving usability issues is needed to address users' needs (Morville 2014). In the case of the event, participants generated ideas that they could implement with their own communities. And because of the diverse backgrounds of the participants, we explored a wide variety of strategies that can be reproduced and/or re-designed easily.

c. Desirable. UX researchers and designers have examined this factor for decades, connecting it to *emotional design*, a concept concerned with prompting users to react positively to products, processes, and content designed with pleasing aesthetics, as perceived by users (Norman 2004; Morville 2014). To this end, the event aimed at providing an experience that was not perceived as just another workshop that Indigenous interpreters and translators had to attend; therefore, it was important to highlight that their role as participants was not to listen to strategies given by scholars of translation and interpreting studies. Their role was to become the experts by ideating the strategies themselves. And the role of the scholars involved in the organization of the event was to moderate their conversations.

d. Accessible. In UX, the accessibility factor refers to the ethical responsibility of designing products, processes, and content that people with disabilities can easily access. In the context of the event and the socioeconomic situation of most Indigenous interpreters and translators, it was evident during the designing stage that funds to attend the event were a socioeconomic deterrent for many participants. Because the great majority of speakers of Indigenous languages in Mexico, and thus Indigenous interpreters, are located in Central and South Mexico, we addressed this issue by situating the conference in the city of Oaxaca and by funding transportation and lodging for those who needed it. Additionally, providing a space for desahogo was also part of making the event accessible because of the trauma associated with Indigenous interpretation.

e. Credible. According to Morville (2014), credibility is a relatively new concept in UX, and it addresses the idea of persuading users to trust particular products, processes, and content enough to use them. It also addresses the ethical implications of using digital technology to persuade users to behave or act in a certain way (e.g., to purchase a product or to use an app). This factor builds on B. J. Fogg's (1998) ideas on *persuasive computers*, which he defines as creating, distributing, or adopting technology that can "change attitudes, beliefs, and behaviors" (385). In the case of our event, albeit planned for a physical space and not for the digital realm, credibility was applied in terms of wanting participants to trust the sources of information by learning from other Indigenous professionals with similar experiences.

f. Valuable. In UX, this factor refers to delivering valuable content, processes, and products by identifying what users actually value. In the case of nonprofit organizations, the user experience must advance the mission of the organization and its users (Morville 2014). To this end, the organizing team envisioned an event that provided participants with takeaways that advance their personal and professional missions. Although our event focused on Indigenous interpreters and translators, the nature of their work as mediators connected other Indigenous and non-Indigenous users to our conversations. Thus, the event produced complex understandings of value, interrogating what is valuable to whom.

g. Equitable. Equity as an indicator that addresses relationships of power is not commonly discussed in UX design. But it became clear right from the start that this was a key factor when working with Indigenous communities. Indigenous people have a distinctive kinesthetic awareness of relationships of power. They have felt it, as Cherrie Moraga and Gloria Anzaldúa (2002) state, in the flesh, since 1492. And thus, we as researchers also had to be aware of these relationships of power during our collaborative planning and during the event so that we could foster equity. During the event, a significant part of our work as facilitators was to ensure that all participants had equal participation and that the strategies proposed fostered equity among Indigenous users and their linguistic practices. Most importantly, because the linguistic rights of Indigenous Peoples are often neglected, once we moved into the minutia of the conversations, participants frequently advocated for anchoring discussions and strategies in international human rights, as indicated in the United Nations Declaration on the Rights of Indigenous Peoples (2007), specifically in Article 13:

1. Indigenous peoples have the right to revitalize, use, develop and transmit to future generations their histories, languages, oral traditions, philosophies, writing systems and literatures, and retain their own names for communities, places and persons.
2. States shall take effective measures to ensure that this right is protected and also to ensure that indigenous peoples can understand and be understood in political, legal and administrative proceedings, where necessary through the provision of interpreters or by other appropriate means. (5)

It should be noted that CEPIADET, the leading Indigenous organization behind the event, is formed by Indigenous attorneys who also provide interpreting and translation services to different communities around the region and beyond. Therefore, they are fully informed of Indigenous rights, as specified in both national and international laws, and of Indigenous rights violations. With that in mind, it was clear that equity would be the main factor from which to build the event and the conversations in it. Grounding the event on equity and human rights was essential to understand the context of the ideas proposed by the participants.

IDEATING

Interpretation and translation services in Indigenous languages are less valued than interpretation and translation services in foreign languages. For that reason, there is a lack of adequate funding provided by public entities to pay for the services of Indigenous interpreters. Nevertheless, partnerships among public institutions and Indigenous organizations have inspired projects that have managed to improve conditions for Indigenous interpreters working in these programs. The following section describes three projects shared by participants through testimonios during the roundtable I moderated. The first project describes a translation endeavor in Peru; the second, an interpretation project in Oaxaca, Mexico; and the third, a translation project in Michoacán, Mexico. These projects served as the foundation of the ideation phase that took place during the roundtable.

Intercultural, Interinstitutional, and Interdisciplinary Translation Project in Peru

Magdalena presented a powerful project that took place in the Peruvian town of Achuar del Pastaza. Multiple institutions became involved in translating official documents for the civil registry to help the Indigenous community of Achuar del Pastaza exercise their civil and citizenship rights in their own language. The organizations involved in this project included the Federación de la Nacionalidad del Pueblo de Achuar del Perú (FENAP) (Citizenship Federation in the Town of Achuar del Perú), the translation and interpretation program at the Universidad Peruana de Ciencias Aplicadas (UPC) (University of Applied Sciences in Peru), and the Instituto Internacional de Derecho y Sociedad (IIDS) (International Institute of Law and Society). As Magdalena explained,

> This was also an interdisciplinary experience because we saw the intercultural lens, the linguistic lens, the translation lens, and the Indigenous rights lens. Each institution contributed something. For example, the intercultural lens came from the academy, from my class; the linguistic lens, from the linguists at the university; the translation lens, from the translators; and the Indigenous rights lens came from the Instituto Internacional de Derecho y Sociedad (IIDS). Then, we signed an

interinstitutional and intercultural agreement because Achuar educators came to work hand in hand with professors at the University.

Involving the Achuar community in this project was the key to accomplishing the outlined objectives. The Indigenous interpreters and translators were recognized and consulted as professionals and experts in the Achuar language and culture in the same way foreign-language experts are recognized and consulted. They created glossaries to reach agreements regarding terms that may not have a translation in Achuar, implemented plain language protocols to avoid obtuse and complicated phrases, and created style guides to facilitate translation. During her presentation, Magdalena explained this process in detail.

> It was very interesting. For example, the certificate said, "By means of this letter, we present to the registrar . . ." with a bombastic language that had the Achuar community all dizzy. What did we do with the pool of linguists? We simplified the document. What does this mean? This means that we simply focused on what was said, who said it, and to whom. That simple. Simplifying the texts became the methodology we found because at the university, no one really knew how to translate the Achuar language. Who was going to say if it was well done or not? Therefore, from the linguistic lens, we only agreed on the methodologies and a bit on the translation process. And because there were certificates that had been translated before, then the terminology was validated for the next documents. We also created protocols. We were looking for quality in the translation. Because there were two translators, one corrected the translation of the other. The protocols consisted of identifying how a particular term was translated and why. Then, glossaries were created because, as I stated before, there were terms like identity, birth, death, and so on. And some of these terms did not exist in the Achuar language. For example, "go to the registrar's office" did not exist in Achuar, then, translating became more complex. Stylesheets were also created to ensure the quality of the translation of certificates. If they added an uppercase letter in one place, then they added it to the whole document. If they added a comma after something, then they had to maintain the same format throughout the document. We wrote notes on the stylesheets and the protocols.

After the documents were simplified, the team focused on the translation process, which was done with the help of Memsource, a software often used by translators to save time during the process. As Magdalena explained, using this software became very helpful during the collaboration between Indigenous and non-Indigenous translators.

> Translators used a program, a software to help the Achuar translators with time management and style, as well as other things. We developed a system in Memsource, which segments the text, allows translators to know how

much is left in the translation, and incorporates terms that have already been translated into the text. With this software, they were able to complete the translation. With an intercultural focus and emphasis on rights, what was most important, I think, is that we built trust. There was communication, and through my contributions, we created a protocol regarding intercultural communication with Indigenous communities through the framework of translation.

Another critical factor in this project was the emphasis on Indigenous rights. This was essential to earn and build trust in alliances between public institutions and Indigenous communities, especially because of the colonizing perspectives that have historically existed within public institutions throughout the continent.

The emphasis on Indigenous rights was really important; in other words, this translation was not intended to perpetuate colonization. This translation was a project for the community to be able to exercise their rights in their own language, which was also really gratifying in terms of strategic partnerships between the academia; the RENIEC, which is a government institution; FENAP, which is an Indigenous organization; an Indigenous community; and also our civil organization through the International Institute (IIDS).... We signed an agreement with an Indigenous rights lens, an agreement where we clearly stated that the university would not be able to appropriate their language. What was translated belonged to the community. We created very clear protocols so that the university would not say, "This is mine, and I'm going to do whatever I want with the Achuar language." No. We protected Indigenous language rights.

For Magdalena, this experience gave the academy the opportunity to learn from Indigenous communities. It allowed public institutions to work not only with a different language but also with a different cosmovision.

What did we learn? I think the most important part of this experience was that it brought institutions together. Imagine! The academia, which only teaches Chinese, French, German, and Italian, suddenly brings Indigenous educators with a different language, a different worldview, and a different cosmovision. Thus, I believe that we should promote alliances between the academia and Native Nations, but always within an Indigenous rights lens that prevents the university from taking their knowledge.... There was also an interchange of knowledges because, for example, the linguists would say, "How are we going to ask them for their password?" because the online resources needed passwords, "How is this going to work? Surely they don't know about passwords." And it turned out that the Achuar educators had two terms for passwords. Why? Because they had worked in Ecuador, and their policies are more advanced over there, so they had two terms. That's an example of how the cultures clashed and how academia was challenged.

This project exemplifies the need to include Indigenous interpreters and translators in the conversations in both TPC and UX as these two fields can clearly help address the communication needs of Indigenous communities. Understanding government documents that prove one's identity in one's own language is a basic need. UX researchers can help provide spaces for interinstitutional, intercultural, and multilingual projects such as this. Language localization and TPC can assist with providing guidance when translating complicated government documents into more manageable texts by localizing translation in Indigenous contexts and cultures and using plain language protocols, as demonstrated by the participants in this project. Also, as Magdalena noticed, we tend to underestimate the digital technical skills of Indigenous people. Thus, UX and TPC can also help by creating digital government content that Indigenous people can easily access through mobile devices.

Young Interpreters and Intercultural Promoters in Oaxaca

Rosa presented a project in which students at the Universidad Autónoma Benito Juárez de Oaxaca (UABJO), through a collaboration with the Centro Profesional Indígena de Asesoría, Defensa y Traducción (CEPIADET), work as Indigenous language interpreters during court hearings. The university considers these interventions as community service hours, and CEPIADET supports these students with training and a small stipend to cover transportation and food costs.

> The Interpreters and Intercultural Promoters program was formed by university students who trained under the leadership of the Centro Profesional Indígena de Asesoría, Defensa y Traducción (CEPIADET) and of the public defenders of the Public Defender's Office, which provided a theoretical and practical training on topics of Indigenous Peoples, human rights, the characteristics of intercultural promoters, translation and interpreting techniques, and topics about the criminal justice system. With this training, interpreters and translators strengthened their abilities, skills, and aptitudes, so they could assist us in helping people and authorities who form the Interinstitutional Committee of People Deprived of Their Freedom. They also supported us with the translation of texts related to access to justice, and everything that had to do with the rights of incarcerated people and Indigenous people.

Rosa explained how this group of young interpreters and intercultural promoters, comprised of Indigenous college students who speak various Indigenous languages, also creates awareness about Indigenous linguistic rights throughout the state of Oaxaca. They build their ethos through statistical research and use technology to disseminate their findings.

> To create awareness about Indigenous topics in the Public Defender's Office, we conducted statistical research about the people we served. Today we know that of the 70,000 cases in the Public Defender's Office, 30,000 concern Indigenous people exclusively. We also know that the Public Defender's Office legally represents 98% of Indigenous people who are deprived of their freedom in different reintegration centers in the entire state, which are 13 in total. . . . We broadcasted our services through videos, videos that were published on the Facebook page of the Public Defender's Office in 16 different languages, that is 16 variants.

Additionally, promoting Indigenous languages among children is central for this group as they believe that involving children is critical not only to creating awareness but also to revitalizing their languages. This group has taken a leadership role in creating awareness through digital technologies. They have translated a few video cartoons into Indigenous languages and are working on acquiring permits to translate more videos that can be disseminated to children from various communities.

> We were able to get some funding at the federal level. These young people in our program are paid 3,600 Mexican pesos each month. This covers the expenses that they incur every weekend. In this way, if there is a hearing or proceeding in an Indigenous language, with this money that these youth are being paid, they can attend and help. Some weeks there may not be a hearing and their work may not be solicited, but this money will still be there to be used by them. Also, for the next trainings, there's going to be a project where they have to go to their communities. We are proposing that they show cartoons, for example, Bugs Bunny from Warner Bros. I have an example in my cell phone where the audio has been replaced with a Zapoteco language. This is one of the ideas we have for these young interpreters and translators to show in their communities to continue revitalizing [Indigenous languages] even more. This is a project directed at kids because children are the ones who are learning.

The group's efforts to create awareness of Indigenous matters have positively affected the region. As Rosa explained, their efforts have prompted the Public Defender's Office to focus on hiring more public defenders with bilingual skills who can help provide better linguistic services for Indigenous people who come in contact with their office.

> Thanks to this visibility, the Public Defender's Office was motivated to prioritize hiring bilingual defenders so that there could be better communication among incarcerated people. To date, we have 21 bilingual defenders who practice in different offices around the state. This, as I mentioned, improves the interaction among incarcerated people and allows us to fulfill our [school's] social service requirement one hundred percent.

Rosa's project invites us to ponder how TPC can help address the needs of Indigenous interpreters working in the legal field.

Understanding interpreting events as technical communication can help create and promote systems that rely on concepts similar to the plain language concepts we use in written texts because, in practice, this is what Indigenous interpreters do when they simplify terms between senders and receivers. Also, TPC can help create and promote processes that allow court interpreters to become acquainted with the rhetorical situation of each case before each interpreting event, in the same way we do with written text. Learning to address the needs of Indigenous court interpreters provides TPC practitioners and UX researchers with an opportunity to become true advocates of users, that is, of all users.

Young Girls as Translation Apprentices in Michoacán

As mentioned in previous chapters, I could only record nine of the testimonios shared by those who participated in my roundtable (mainly because of technical issues). Therefore, only nine shared testimonios were analyzed in the previous chapters. Still, I believe another significant project should also be discussed in this section; the project presented by Luz, a P'urhepecha[1] translator from Michoacán who works with an Indigenous organization that focuses on teaching young girls to translate comic books from Spanish to their mother tongue. To convey the information about this project, I rely on my notes and the material the presenter shared. Although I could not record Luz's testimonio, I recorded her introduction: "My name is [Luz], and I'm from Michoacán, from a town called Santa Fe de la Laguna. I speak P'urhepecha, and we are working on a project called *Majkuksï Monharitantani Karákatechani*, which means *Traduciendo Juntas* [Translating Together]. In this project, we're translating comics and solving several [writing and translation] issues."

The project of which Luz is part aims to promote the culture of writing as a means to revitalize and preserve the P'urhepecha language. It is an initiative co-produced in collaboration with Hormiguera Editorial and the Escuela Nacional de Estudios Superiores (ENES) Campus Morelia, a site of the Universidad Nacional Autónoma de México (UNAM) in Morelia, Michoacán, Mexico (see figure 6.5). Colectivo Uantakua was formed by ten young girls from six to twenty-eight years old who are learning translation techniques (Luz 2019).

The Twitter post by ENES shown in figure 6.5 is an example of how the academia became involved in promoting not only the P'urhepecha language but also Indigenous collective initiatives through the project

142   IDEATING AND RE-DESIGNING

Figure 6.5. Image shared on Twitter by ENES Morelia promoting the project Traduciendo Juntas (ENES 2019)

Traduciendo Juntas (Translating Together). In their message, ENES calls for the help of typographers to join the project, and on the right side of the post, they show some of their collective work.

The group gets together once a month to provide workshops on translation to the girls involved in the different projects. As in Magdalena's translation project, during their workshops they teach girls to identify the basic information of what is translated, when, and where. They also work on strategies to create glossaries and edit what is written, always considering the girls' opinions as they are the native speakers of the language in which the translations are done. They work on first drafts, revisions, and final drafts. Luz also explained that the university works on the comic illustrations while the girls help translate.

Figure 6.6 shows one of the comic strips translated by the young P'urhepecha girls who participate in the Colectivo Uantakua (Traduciendo Juntas). During the roundtable I moderated, Luz showed us several examples. The objectives of the project, as described by Luz, in no order of importance, are as follows:

ERA, TATA PEGOSTE, URHEPIA T'U

[Comic panel with speech: "Ia." / "PSSSH PSSH" / "Noteruche meni jamaka tantani unichani."]

Figure 6.6. Comic strip translated by Colectivo Uantakua (Luz 2019)

1. Recuperate P'urhepecha words that people have stopped using
2. Learn about editing and publishing
3. Relearn to value the P'urhepecha language
4. Recognize translation as a practice that requires patience, discipline, commitment, and responsibility

Luz also shared the challenges that this initiative faces:

1. Continue translating
2. Work on the publication of translations already co-created
3. Create international awareness about the P'urhepecha language
4. Add more P'urhepecha speakers to language revitalization efforts
5. Show the P'urhepecha community what a comic is
6. Promote literacy in the P'urhepecha language

Projects such as this should also be addressed as technical communication, not only because creating glossaries, editing, revising, and translating are all part of the work of technical writers but also because this work is *not* done with the main purpose of entertaining, as in the objective of the original comic book. Translation projects involving Indigenous languages have a conscious activist purpose that places Indigenous languages and translation as practices of linguistic activism. Therefore, both UX and TPC can also be of tremendous help to

Indigenous organizations working on translating children's texts or video cartoons from Western languages to Indigenous languages, like the projects in which Rosa is involved.

As seen in this section, Indigenous and non-Indigenous translators and interpreters have worked together on different projects to create awareness about Indigenous languages through translation and interpretation initiatives that involve technical and professional communication skills. Magdalena's description of how plain language was used within an Indigenous language context; Rosa's explanation of how Indigenous college students get involved in the technicalities of interpreting legal terms, translating for video projects, and revitalizing Indigenous languages through digital spaces; and Luz's explanation of how young P'urhepecha girls collaborate with illustrators to create comics in order to revitalize her language are without a doubt examples of Indigenous interpreters and translators as technical communicators with agency who advocate for the promotion and advancement of Indigenous languages.

(RE)DESIGNING

Through the conversations at the roundtable I moderated, participants identified several strategies to help improve employment conditions and opportunities for Indigenous interpreters and translators in the public sector. As a roundtable, we identified strategies that can be launched from their own communities, including:

1. Creating a multinational master list of Indigenous interpreters
2. Making better use of emerging technologies (e.g., videos, documentaries, social media sites) to develop materials with information regarding
    a. associations that Indigenous interpreters can join
    b. how to manage interpretation in different professional contexts (medical, legal, and educational)
    c. how to navigate various sociopolitical contexts in the country/ state where interpretation and translation services are rendered
3. Establishing methodologies for evaluating interpretation and translation practices from the perspective of Indigenous interpreters, including developing
    a. linguistic standards
    b. cultural standards that recognize the interpreter as a cultural mediator and that recognize Indigenous interpretation as an activity that cannot be impartial
    c. professional standards that allow interpreters to render high-quality services in their respective sectors (legal, medical, educational)

4. Partnering with international academic institutions with access to funding that can support training for Indigenous interpreters and that can help develop training materials for Indigenous interpreters

Our roundtable also generated general strategies to foster self-reliance and sovereignty for Indigenous interpreters:

1. Make more effective use of language rights laws and of the role that human rights organizations can play in enforcing these rights
2. Recognize the legal pluralism of Indigenous communities in discussions about rights
3. Train Indigenous interpreters in the legal rights of Indigenous communities within international human rights standards
4. Advocate for the recognition of Indigenous interpretation as an activity that is *not neutral*
5. Advocate for the recognition of the interpreter as a cultural mediator
6. Develop an awareness campaign for law enforcement officers

The shared experiences and challenges and the possibilities identified in the roundtable conversations push for a conceptualization of Indigenous translation and interpretation that does not align with the Eurocentric perspective of interpretation as a practice devoid of historical, cultural, and social contexts. Indigenous interpreters advocate for better compensation within the public sector. Indigenous interpreters advocate for being recognized as cultural and social mediators. Indigenous interpreters consistently advocate for their languages to be valued. And above all, as Magdalena explained, Indigenous language interpreters and translators advocate for Indigenous communities to exercise their rights in their own languages.

Both UX and TPC can play an active role in designing better experiences for both Indigenous professionals and Indigenous users. For example, TPC programs can start by forming long-term alliances with Indigenous organizations to mitigate the disjunctions discussed in this section. Languages and cultures are not barriers when differences are acknowledged and respected. Indigenous organizations know this as they have always worked within this context—because there are hundreds of Indigenous languages and cultures throughout the Americas, and TPC can learn from their multicultural and multilingual approaches.

UX researchers can also help Indigenous communities by learning to address their needs as seen through their own cosmovision, as this work attempts to do. Whereas the issues affecting Indigenous communities were and continue to be triggered by governments throughout the

Americas, all of us non-Indigenous people living in native spaces are indebted to Indigenous communities. Academia in particular can bear some of the responsibility of repaying this debt by addressing the professionalization needs of Indigenous professionals.

All three projects discussed here prove that alliances between Indigenous communities and public institutions are vital to promoting awareness about Indigenous matters, revitalizing Indigenous languages, and professionalizing the field of Indigenous interpreters and translators. In the conclusion of this book, I examine important implications of the topics analyzed throughout this study and provide more guidance about how TPC and UX can better assist Indigenous users through Indigenous approaches to user experience.

CONCLUSION

Technical and professional communication (TPC) and translation and interpreting studies (TIS) have focused on analyzing traditions that disenfranchise practices that Indigenous communities have maintained for more than one thousand years. Although TIS inherently analyzes translation and interpreting from multilingual perspectives, there is a clear disconnect between contemporary researchers who advocate for methods that challenge neutrality and invisibility ideologies, educators who continue to support outdated practices masked behind a so-called *code of ethics*, and translators and interpreters in the field strained by the emotional stress caused by outdated codes of ethics and notions of neutrality. While TPC has supported community-based research and social justice approaches for decades, Indigenous practices are still sidelined in this field. Even user-centered approaches and language-localization research in TPC, which often address translation, rarely discuss Indigenous translation and interpreting practices. TPC scholars who work with Indigenous groups from Latin America can be counted on one hand. This study aimed at addressing the gaps in these fields through an Indigenous approach to user experience research.

IMPLICATIONS OF THE STUDY

The interdisciplinary nature of my work has several implications for Indigenous practitioners, researchers, and educators. First, let me be clear, Indigenous interpreters and translators are well aware of the issues they face in relation to public institutions, and they do not need user experience models or processes to point out these issues. Notwithstanding, the primary goal of this study is to help create better, more equitable systems and processes that directly benefit Indigenous people who come in contact with Western public systems. My work can be used to design better experiences for Indigenous interpreters and translators so they can, in turn, provide better linguistic services for Indigenous users.

Design thinking is not a static method, as I initially assumed, but an influx process of co-constructing and re-designing, as we all come to it with knowledge to contribute, then learn new knowledge during the process to subsequently do something with it. We must consider equity as a core factor in UX because there are always relationships of power involved in designing products, content, and processes. While UX designers might not be aware of this, users of underrepresented groups are (and accessibility does not address relationships of power). Fostering equity in UX means adjusting approaches to ensure that all participants are considered. In UX, people's contexts matter a lot, and it is important to be comfortable with being uncomfortable with ambiguity, especially when working with multicultural and multilingual groups. We cannot always approach UX and/or design thinking in the same way. Prototyping is not necessarily a stage that would yield an immediate tangible product or process when working with Indigenous people. We all understand the process of designing differently. For Indigenous people, as in the case of many underrepresented cultures, this process is an exceptionally collective act (like cooking as a family in many Latinx cultures).

This project offers an Indigenous approach to UX research that helps incorporate diverse methods because Western practices do not always address the needs of all users. Utilizing circumscribed UX methods that include concepts foreign to the contexts of underrepresented communities may hinder communication and trust between the participants and the researcher(s) during a research project. Therefore, testimonios can be a valuable tool to design UX projects with Indigenous individuals and other underrepresented groups who express their needs as collective needs. Testimonios intrinsically examine a user's experience in the social and cultural context of a group while revealing the user's civic engagement and call to action through narrative and dialogue, increasing user agency and supporting user advocacy. This method works well with *superdiversified* (Cardinal 2022) groups with different perspectives as it helps practitioners understand important cultural differences when grounding research on cultural contexts. When conducting research with Indigenous groups specifically, researchers must consider the community-oriented roles of Indigenous individuals. Each participant's personal and collective narratives and the dialogue embedded in these narratives prompt participants to reflect on issues at a deeper level, engaging with their unique differences while revealing general similarities.

To work with testimonios as a methodology, however, practitioners must invest time to build a relationship with the participants. It is not a methodology that can be done through impersonal modes (like email). Researchers must work with participants in person or through a digital video conferencing tool that allows participants to interact with researchers and with one another. Most importantly, practitioners must understand this methodology from an Indigenous lens (Medina 2018; Menchú 1984; Smith 2012), especially when working with the unique elements of desahogo and dialogue.

> From a Western lens, a narrative from a collective voice might seem out of place; a dialogical approach to research where researchers get involved in the conversations might seem biased; and an oral cathartic release of emotion in highly technical environments often seems unprofessional. From an Indigenous lens, however, the collective voice of a narrative points out the social and cultural roots and effects of complex issues; a dialogical approach engages participants in a deep reflection and helps negotiate meaning in oral interactions, balancing the relationships of power in research; and an emotional desahogo leads participants to a call to action that empowers users during a research project. (Rivera 2022b, 24)

Further, this study fosters an antioppressive pedagogy that advocates for the agency of Indigenous interpreters and translators. TIS cannot teach translation and interpretation the same way across all languages if equity is the goal. Indigenous languages have a historical and political context that needs to be addressed differently. Indigenous interpretation in particular entails mediating linguistic nuances, technical terminology, cultures from different worldviews, institutional regulations versus traditional customs, and sociopolitical environments, all while negotiating values, loyalties, emotions, relationships of power, biases, and dispositions of everyone involved in an interpreting event. Acknowledging that Indigenous translation and interpretation practices are ambiguous and never unbiased is crucial to create spaces that foster Indigenous interpreting strategies, such as dialogue and desahogo. Additionally, as an interdisciplinary field that aims at improving processes and clarifying specialized terminology, often in translation through language-localization strategies, TPC's collaborations with TIS should be more conspicuous.

This study can also help promote antioppressive and antiracist pedagogies in TPC and rhetoric and composition courses—which rely heavily on Eurocentric curricula—to foster better learning experiences for multicultural and multilingual students who struggle to connect with curricula extraneous to their backgrounds. What if we change the UI model I presented in chapter 6 and place educators instead of Indigenous interpreters at the center? Educators are also technical communicators,

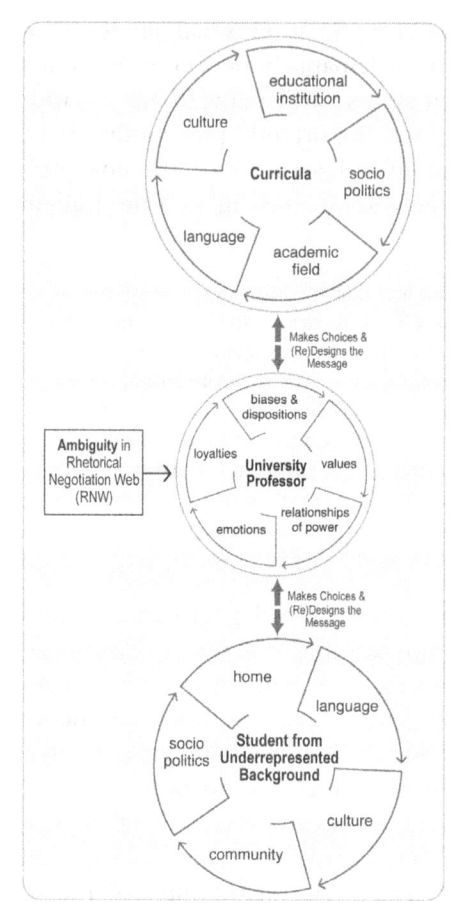

Figure 7.1. The user interface of university classrooms

for we teach and clarify highly specialized content in our classrooms. On the top of the UI of university classrooms we place the curriculum in our classrooms, and on the bottom we place students from underrepresented communities (see figure 7.1). These students navigate an interface imbued in cultural, educational, sociopolitical, and linguistic contexts different from their homes and communities. Not to mention that, in most cases, students from underrepresented backgrounds take university courses from someone more aligned with the Eurocentric perspectives of our educational systems.

While exposing students to material written by authors who belong to underrepresented groups is not enough, it can be the factor that engages them. Curricula that address issues of race, ethnicity, gender, sexuality, citizenship, and language from the lens of authors who experience them, not as an afterthought lesson but as the core of a course, engage students who feel disconnected and often become the reason why these students stay in academia. Students who can hide their nondominant backgrounds, for lack of a better word, would do so to be accepted in highly hierarchical educational spaces, as Carlos commented in his testimonio (chapter 4, table 4.2). Diverse curricula with diverse pedagogical strategies that foster community building, collaboration, and multimodality—curricula that understand multicultural and multilingual backgrounds as assets rather than deficits—benefit all students, preparing them for careers in the superdiversified societies of our everyday life today. What rhetorical choices are we making to re-design the messages we transmit to our students? What kind of feedback do we receive from students from underrepresented backgrounds to continuously re-design our curricula? Or is our classroom UI a one-way street? This study invites educators from all backgrounds to inquire about the various ways we can help balance the disparities of the institutional verticality embedded in our university classrooms.

Lastly, this study can also lead to writing-with-the-community projects such as the unconference proceedings written under the leadership of CEPIADET, the Indigenous NGO behind the event I cite here, and in collaboration with Indigenous translators, TPC, and TIS scholars from Mexico, the United States, and Canada (Castellanos García et al. 2022). The proceedings contain translations in five different languages: Mixe, Zapoteco, Mixteco, Spanish, and English.

FINAL THOUGHTS

I hope that this book can help produce methods and systems that better address the long-term needs of Indigenous interpreters and translators so that they can better support their communities. This book also emphasizes the need for Mestize, Latinx, and Hispanic scholars to grapple with and work through our own relationships with indigeneity in ethical ways. It is wrong and unfair that Indigenous organizations carry the heavy load of Indigenous advocacy. There is much more that can be done and should be done by those of us who inhabit Indigenous lands, using Western systems and speaking Western languages effortlessly.

NOTES

INTRODUCTION

1. Although I use the terms "Nahua" and "Nahuatl" to refer to one of the main national languages of Mexico, other scholars writing in English might use the terms "Náhuatl," "Nahual," or "Nahuat." The difference rests on the group category to which one refers. According to the *Catálogo de las Lenguas Indígenas Nacionales*, published by the Instituto Nacional de Lenguas Indígenas (INALI 2008), there are eleven linguistic families in Mexico, one of which is the Yuto-Nahua. Then, there are sixty-eight linguistic groups, one of which is the Náhuatl. There are also 364 linguistic variants, all of which should be treated as autonomous languages. The Náhuatl linguistic group alone has thirty variants, of which some self-denominate in different ways, including Nauta, Tla'tol, Masehuatl Tajtol, Náhuatl, Mexcatl, Nahuat, and Mexicano, to name a few. Throughout the book, I use Nahuatl and Nahua, from the Yuto-Nahua linguistic family, with the intention of including as many categories as possible.
2. The term "cosmovision," or worldview, is often associated with Mesoamerican cultures. Nonetheless, it is a concept that can be applied to any society. Richard DeWitt (2013) defines the term "cosmovision" as "a system of beliefs interconnected like the pieces of a puzzle. In other words, a cosmovision is not merely a collection of separate independent, unrelated beliefs, but a system of beliefs that are intertwined, interrelated, and interconnected" (19).
3. In an unconference format, participants drive the content of the event in order to disrupt hierarchical roles. In our event, the attendants became the presenters, and my academic colleagues and I moderated their contributions.
4. The original terms used to categorize individuals of mixed Spanish and Indigenous heritage were "Mestizo" (for male) and "Mestiza" (for female). "Mestiz@," "Mestizx," and "Mestize" are gender-inclusive neologisms (Rivera 2022b, 10). I use "Mestize" as a gender-inclusive neologism in this book. It should be noted that Latin Americans are very diverse. Although this section addresses Indigenous and Mestize terms, these are not the only identities Latin Americans embody. People also self-identify as Afro Descendants, Latinasians, and Whites, to name a few.
5. The Global South is, in simple terms, a critical concept that addresses inequalities and race through geographic lenses.
6. Although user-centered design (UCD), user experience (UX), and language localization are interconnected, there are significant differences. Kirstin Hierholzer (2013) defines UCD as a "philosophy that puts the user in the center of the design and development process." Localization is a concept used often in UX to address translation. Laura Gonzales and Rebecca Zantjer (2015) define localization as "the process of adapting content for a specific culture" (272). In other words, localizing is translating but with a greater consideration for culture and context, and it is not necessarily done word for word (Gonzales and Zantjer 2015). Moreover, this study understands UX as an interdisciplinary research methodology that focuses on the users, what they need, and what they value with the purpose of designing better, more usable products and more effective content and processes.

CHAPTER 1: INTERSECTING THEORIES AND DISCIPLINES

1. Because both the "post-" and the "de-" resist coloniality and expose its effects, it is difficult to discuss one without the other—so much so that theorists such as Gloria Anzaldúa and Walter Mignolo are often associated with both postcolonial theory (because of the ideas of the era in which these authors became known) and decolonial theory (because of the place from where they write) (Rivera 2020).
2. The words "humana" and "humano" in Spanish mean both human and humane, and Vasconcelos evidently refers to the latter in his original text.
3. "La Raza Cósmica" was published about a decade after the Mexican Revolution took place. Without undermining the complexities of the Mexican Revolution, this upheaval was the result of the thirty-year dictatorship of Porfirio Díaz, whose dogma consisted of promoting *latifundios* (large plantation-style estates) that only enriched the rich and neglected the needs of the poor. Díaz's government was strongly influenced by foreign ideas; therefore, during the decade after the Mexican Revolution, the government was swayed toward a nationalistic view that placed the Mexican identity at its core (Rivera 2020). In a nutshell, nationalism in Mexico during the decades after the Mexican Revolution meant connecting with Indigenous roots and placing agrarian and labor workers at the forefront of the political agenda. It also meant homogenizing Mexico under the Mestize umbrella and the Spanish language.

CHAPTER 2: DESIGNING THE RESEARCH

1. "Chicanos," "Chicanas," and "Chicanxs" are terms used by individuals of Mexican heritage in the United States. Some Chicanxs also self-identify as Indigenous and/or Mestizes. Not all Mexican Americans see themselves as Chicanxs.
2. The complexity of the issues and the strategies proposed to solve these issues made it difficult for participants to test ideas on the spot. In this two-day event, for example, participants could not simply take a strategy and drive to a court hearing to test it.

CHAPTER 3: EMPATHIZING

1. In Peru, Felipillo is a figure similar to Malintzin in Mexico (a.k.a. Doña Marina and La Malinche). La Malinche was an Indigenous woman who interpreted for Hernán Cortés in two Indigenous languages, Nahuatl and another Indigenous language from the state of Tabasco in Mexico. Although her figure for many Chicanas has become a feminist symbol, her name in Mexico is still widely linked to Indigenous treachery.

CHAPTER 5: SYNTHESIZING NEEDS AND ISSUES

1. CERESO stands for Centro de Reinserción Social (Social Reintegration Center). Antonia coordinates a group of Tarahumara court interpreters—mostly women—who provide services in the CERESO of Chihuahua City in Mexico.

CHAPTER 6: IDEATING AND RE-DESIGNING

1. As in the case of most Indigenous languages, P'urhepecha has a Spanish spelling and an Indigenous self-denomination (the way in which P'urhepecha speakers write it). In Spanish, it is written as "Purépecha," but its self-denomination is "P'urhepecha." The P'urhepecha language is a variant of the Tarasco linguistic group (INALI 2008).

REFERENCES

Agboka, Godwin. 2014. "Decolonial Methodologies: Social Justice Perspectives in Intercultural Technical Communication Research." *Journal of Technical Writing and Communication* 44 (3): 297–327.

Alonso, Icíar, and Gertrudis Payás. 2008. "Sobre Alfaqueques y Nahuatlalos: Nuevas Aportaciones a la Historia de la Interpretación." In *Investigación y Práctica en Traducción e Interpretación en los Servicios Públicos: Desafíos y Alianzas*, edited by Carmen Valero-Garcés, Carmen Pena Díaz, and Raquel Lázaro Gutiérrez, 38–51. Madrid: Editorial Universidad de Alcalá.

Angelelli, Claudia. 2004. *Revisiting the Interpreter's Role: A Study of Conference, Court, and Medical Interpreters in Canada, Mexico, and the United States*. Amsterdam: John Benjamins Publishing.

Angelelli, Claudia V. 2016. "Looking Back: A Study of (Ad-Hoc) Family Interpreters." *European Journal of Applied Linguistics* 4 (1): 5–31.

Angelelli, Claudia V., and Brian James Baer. 2016. *Researching Translation and Interpreting*. New York: Routledge.

Anzaldúa, Gloria. 2012. *Borderlands, La Frontera: The New Mestiza*, 4th ed. San Francisco: Aunt Lute Books.

Arias, Arturo. 2016. "New Indigenous Literatures in the Making: A Contribution to Decoloniality." In *Decolonial Approaches to Latin American Literatures and Cultures*, edited by Juan G. Ramos and Tara Daly, 77–95. London: Palgrave Macmillan.

Baca, Damián. 2009. "Rethinking Composition, Five Hundred Years Later." *JAC* 29 (1/2): 229–242.

Benmayor, Rina. 2012. "Digital Testimonio as a Signature Pedagogy for Latin@ Studies." *Equity and Excellence in Education* 45 (3): 507–524.

Bhabha, Homi. 1984. "Of Mimicry and Man: The Ambivalence of Colonial Discourse." *Discipleship: A Special Issue on Psychoanalysis* 28: 125–133.

Bhabha, Homi. 1994. *The Location of Culture*. New York: Routledge.

Biernacka, Agnieszka. 2008. "Intérprete en el Contexto Sociológico de la Sala de Audiencias." In *Investigación y Práctica en Traducción e Interpretación en los Servicios Públicos: Desafíos y Alianzas*, edited by Carmen Valero-Garcés, Carmen Pena Díaz, and Raquel Lázaro Gutiérrez, 298–313. Madrid: Universidad de Alcalá.

Blanco, Sergio. 2018. "Justice in Translation." *New York Times*, December 10, 2018. https://www.nytimes.com/video/opinion/100000006194836/mexico-justice-system.html?searchResultPosition=1.

Cardinal, Alison. 2022. "Superdiversity: An Audience Analysis Praxis for Enacting Social Justice in Technical Communication." *Technical Communication Quarterly* 31 (4): 343–355. https://doi.org/10.1080/10572252.2022.2056637.

Castañeda, Antonia. 1998. "Language and Other Lethal Weapons: Cultural Politics and the Rites of Children as Translators of Culture." *Chicana/o Latina/o Law Review* 19 (1): 229–242.

Castellanos García, Abigail. 2021. "El Fomento de las Lenguas Indígenas en los Espacios Públicos." Paper presented at the annual convention of the Latin American Studies Association, Virtual, May 2021.

REFERENCES

Castellanos García, Abigail, Laura Gonzales, Cristina V. Kleinert, Tomás López Sarabia, Edith Matías Juan, Mónica Morales-Good, and Nora K. Rivera. 2022. *Indigenous Language Interpreters and Translators: Toward the Full Enactment of All Language Rights.* Lexington, KY: Intermezzo. https://intermezzo.enculturation.net/16-gonzales-et-al.htm?fbclid=IwAR3-4coyNo_i1M-wP3td4vXo8LSPiTmpCBrU5cUmPyp93RhwFqVyfcnW-YM.

CEPIADET A. C. n.d. "Home." YouTube. Accessed September 8, 2020. https://www.youtube.com/c/CEPIADETAC/videos.

Ceribelli, Alessandra. 2013. "Relación de las Cosas de Yucatán de Fray Diego De Landa: Una Mirada Europea sobre la Realidad Americana." *Cuadernos de Aleph* 5: 39–55.

Chávez Leyva, Yolanda. 2003. "In Ixtli in Yóllotl/A Face and a Heart: Listening to the Ancestors." *Studies in American Indian Literatures* 15 (3/4): 96–127.

Delgado, L. Elena, Rolando J. Romero, and Walter Mignolo. 2000. "Local Histories and Global Designs: An Interview with Walter Mignolo." *Discourse* 22 (3): 7–33.

DeWitt, Richard. 2013. *Cosmovisiones. Una Introducción a la Historia y a la Filosofía de la Ciencia.* Vilassar de Dalt, Spain: Biblioteca Buridán.

Díaz, Gisele, and Alan Rodgers. 1993. *The Codex Borgia: A Full-Color Restoration of the Ancient Mexican Manuscript.* Mineola, NY: Dover.

Dorpenyo, Isidore K. 2020. "Decolonial Methodology as a Framework for Localization and Social Justice Study in Resource-Mismanaged Context." In *User Localization Strategies in the Face of Technological Breakdown: Biometric in Ghana's Elections*, 53–78. New York: Palgrave Macmillan.

Driskill, Qwo-Li. 2015. "Decolonial Skillshares: Indigenous Rhetorics as Radical Practice." In *Survivance, Sovereignty, and Story: Teaching American Indian Rhetorics*, edited by Lisa King, Rose Gubele, and Joyce Rain Anderson, 57–78. Logan: Utah State University Press.

Durá, Lucía. 2015. "~~What's Wrong Here?~~ What's Right Here? Introducing the Positive Deviance Approach to Community-Based Work." *Connexions: International Professional Communication Journal* 4 (1): 57–89.

Durá, Lucía, Laura Gonzales, and Guillermina Solis. 2019. "Creating a Bilingual, Localized Glossary for End-of-Life Decision-Making." *SIGDOC'19 Proceedings of the 37th ACM International Conference on the Design of Communication*, Article 30.

Echeverría, Bolívar. 2010. *Modernidad y Blanquitud.* Mexico City: Ediciones Era.

Escuela Nacional de Estudios Superiores (ENES Unidad Morelia) (@ENESMoreliaUNAM). 2019. "#ComunidadENES; hacemos extensiva la invitación del Laboratorio de Publicaciones al taller: Majkuksï monharitantani karákatechani / Traduciendo juntas. (Traducción de historietas al purépecha) 21 de febrero a las 11am. Informes en el correo: cgarduno@enesmorelia.unam.mx." Twitter, February 15, 2019, 5:32 p.m. https://twitter.com/ENESMoreliaUNAM/status/1096567990382940160.

First Nations Technology Council. n.d. "UX Design." First Nations Technology Council. Accessed September 8, 2020. https://technologycouncil.ca/projects/ux-design/.

Flores, Judith, and Silvia Garcia. 2009. "Latina Testimonios: A Reflexive, Critical Analysis of a 'Latina Space' at a Predominantly White Campus." *Race Ethnicity and Education* 12 (2): 155–172.

Fogg, B. J. 1998. "Captology: The Study of Computers as Persuasive Technologies." In *CHI 98 Conference Summary on Human Factors in Computing Systems (CHI '98)*, 385. New York: Association for Computing Machinery.

Gentile, Paola. 2014. "The Conflict between the Interpreter's Role and Professional Status: A Sociological Perspective." In *(Re)Visiting Ethics and Ideology in Situations of Conflicts*, edited by Carmen Valero-Garcés, 195–204. Madrid: Universidad de Alcalá.

Giddens, Anthony. 1984. *The Constitution of Society.* Berkeley: University of California Press.

Golitsis, Pantelis. 2021. "Aristotle on Ambiguity." In *Strategies of Ambiguity in Ancient Literature*, edited by Martin Vöhler, Therese Fuhrer, and Stavros Frangoulidis, 11–28. Berlin: De Gruyter.

Gonzales, Laura. 2018. *Sites of Translation: What Multilinguals Can Teach Us about Digital Writing and Rhetoric.* Ann Arbor: University of Michigan Press.

Gonzales, Laura. 2021. "(Re)Framing Multilingual Technical Communication with Indigenous Language Interpreters and Translators." *Technical Communication Quarterly* 31 (1): 1–16.

Gonzales, Laura, Kendall Leon, and Ann Shivers-McNair. 2020. "Testimonios from Faculty Developing Technical and Professional Writing Programs at Hispanic-Serving Institutions." *Programmatic Perspectives* 11 (2): 67–93.

Gonzales, Laura, Robin Lewy, Erika Hernández Cuevas, and Vianna Lucía González Ajiataz. 2022. "(Re)Designing Technical Documentation about COVID-19 with and for Indigenous Communities in Gainesville, Florida, Oaxaca De Juárez, Mexico, and Quetzaltenango, Guatemala." *IEEE Transactions on Professional Communication* 65 (1): 34–49.

Gonzales, Laura, and Rebecca Zantjer. 2015. "Translation as a User-Localization Practice." *Technical Communication* 62 (4): 271–284.

Haas, Angela M. 2007. "Wampum as Hypertext: An American Indian Intellectual Tradition of Multimedia Theory and Practice." *Studies in American Indian Literatures* 19 (4): 77–100.

Haas, Angela M. 2012. "Race, Rhetoric, and Technology: A Case Study of Decolonial Technical Communication Theory, Methodology, and Pedagogy." *Journal of Business and Technical Communication* 26 (3): 277–310.

Haas, Angela. 2015. "Towards a Decolonial Digital and Visual American Indian Rhetorics Pedagogy." In *Survivance, Sovereignty, and Story: Teaching American Indian Rhetorics,* edited by Lisa King, Rose Gubele, and Joyce Rain Anderson, 188–208. Logan: Utah State University Press.

Hierholzer, Kirstin. 2013. "UCD, UX, Usability—So What's the Difference?" *User Experience at the University of Oregon* (blog), *University of Oregon.* September 20, 2013. https://blogs.uoregon.edu/uxuo/2013/09/20/ucd-ux-usability-so-whats-the-difference/.

Houston, Stephen, John Robertson, and David Stuart. 2000. "The Language of Classic Maya Inscriptions." *Current Anthropology* 41 (3): 321–356.

Inclán Solís, Daniel. 2016. "Against Ventriloquism: Notes on the Uses and Misuses of the Translation of the Subaltern Knowledge in Latin America." *Cultura Hombre y Sociedad* 26 (1): 61–80.

Inghilleri, Moira. 2012. *Interpreting Justice: Ethics, Politics, and Language.* New York: Routledge.

Instituto Nacional de Lenguas Indígenas (INALI). 2008. *Catálogo de las Lenguas Indígenas Nacionales.* Diario Oficial de la Federación.

Itchuaqiyaq, Cana Uluak, and Breanne Matheson. 2021. "Decolonizing Decoloniality: Considering the (Mis)Use of Decolonial Frameworks in TPC Scholarship." *Communication Design Quarterly Online First* 9 (1): 20–31. https://sigdoc.acm.org/wp-content/uploads/2021/02/CDQ_20007_Itchuaqiyaq_Matheson.pdf.

Jawetz, Tom, and Scott Shuchart. 2019. *Language Access Has Life-or-Death Consequences for Migrants.* Washington, D.C.: Center for American Progress. https://cdn.americanprogress.org/content/uploads/2019/02/21060810/Language-Access-for-Migrants-7.pdf.

Jones, Natasha N. 2016. "The Technical Communicator as Advocate: Integrating a Social Justice Approach in Technical Communication." *Journal of Technical Writing and Communication* 46 (3): 1–20.

Jones, Natasha N., Kristen R. Moore, and Rebecca Walton. 2016. "Disrupting the Past to Disrupt the Future: An Antenarrative of Technical Communication." *Technical Communication Quarterly* 25 (4): 211–229.

King, Thomas. 2003. *The Truth about Stories: A Native Narrative.* Minneapolis: University of Minnesota Press.

Kleinert, Cristina. 2015. "La Formación de Intérpretes de Lenguas Indígenas para la Justiciar en México: Sociología de las Ausencias y Agencia Decolonial." *Sendebar* 26: 235–654.
Kleinert, Cristina. 2016. Didáctica para la Formación de Intérpretes en Lenguas Nacionales de México: Trabajar de Manera Multilingüe. *Entreculturas* 7 (8): 599–623.
Latour, Bruno. 2005. *Reassembling the Social: An Introduction to Actor-Network Theory*. Oxford: Oxford University Press.
León-Portilla, Miguel. 1991. *Huehuehtlahtolli: Testimonios de la Antigua Palabra*. Mexico, D.F.: Fondo de Cultura Económica.
León-Portilla, Miguel. 2010. *Los Antiguos Mexicanos a Través de Sus Crónicas y Cantares* [eBook edition]. Mexico, D.F.: Fondo de Cultura Económica.
Lugones, María C., and Elizabeth V. Spelman. 1983. "Have We Got a Theory for You! Feminist Theory, Cultural Imperialism and the Demand for 'The Woman's Voice.'" *Women's Studies International Forum* 6 (6): 573–581.
Luhmann, Niklas. 1996. *Social Systems: Writing Science*. Translated by John Bednarz Jr. and Dirk Baecker. Redwood City, CA: Stanford University Press.
Lunsford, Andrea A. 1998. "Toward a Mestiza Rhetoric: Gloria Anzaldúa on Composition and Postcoloniality." *The Journal of Advanced Composition* 18 (1): 1–27.
Luz (pseudonym). 2019. "Working Together: Colectivo Uantakua." PowerPoint slides presented at the First International Conference for Indigenous Interpreters and Translators, Oaxaca, Mexico, August 2019.
Lyons, Scott Richard. 2000. "Rhetorical Sovereignty: What Do American Indians Want from Writing?" *College Composition and Communication* 51 (3): 447–468.
Macri, Martha J. 2005. "Nahua Loan Words from the Early Classic Period: Words for Cacao Preparation on a Río Azul Ceramic Vessel." *Ancient Mesoamerica* 16: 321–326.
Matías Juan, Edith. 2021. "La Formación de Intérpretes Indígenas como un Derecho Humano." Paper presented at the annual convention of the Latin American Studies Association, Virtual, May 2021.
Medina, Cruz. 2018. "Digital Latin@ Storytelling: Testimonio as Multi-Modal Resistance." In *Racial Shorthand: Coded Discrimination Contested in Social Media*, edited by Cruz Medina and Octavio Pimentel. Logan: Computers and Composition Digital Press/Utah State University Press. http://ccdigitalpress.org/book/shorthand/chapter_medina.html.
Menchú, Rigoberta. 1984. *I, Rigoberta Menchú: An Indian Woman in Guatemala*. Edited by Elisabeth Burgos-Debray. Translated by Ann Wright. London: Verso.
Mignolo, Walter. 1992. "The Darker Side of the Renaissance: Colonization and the Discontinuity of the Classical Tradition." *Renaissance Quarterly* 45 (4): 808–828.
Mignolo, Walter. 2003. *The Darker Side of the Renaissance: Literacy, Territoriality, and Colonization*, 2nd ed. Ann Arbor: University of Michigan Press.
Mignolo, Walter. 2005. *The Idea of Latin America*. Hoboken, NJ: Blackwell Publishing.
Mignolo, Walter. 2009. "Epistemic Disobedience, Independent Thought and De-Colonial Freedom." *Theory, Culture, and Society* 26 (7–8): 1–23.
Mora Curriao, Maribel. 2007. "La Construcción de Sí Mismo en Testimonios de Dos Indígenas Contemporáneos." *Documentos Lingüísticos y Literarios*, no. 30. http://www.revistadll.cl/index.php/revistadll/article/view/214.
Moraga, Cherríe, and Gloria Anzaldúa, eds. 2002. *This Bridge Called My Back: Writings by Radical Women of Color*, 3rd ed. Berkeley, CA: Third Woman Press.
Morales-Good, Mónica. 2020. "Voiceless Voices in a Silent Zone: The Role of the Indigenous Language Interpreter in Oaxaca, Mexico." Dissertation, University of British Columbia. UBC Theses and Dissertations Open Collections.
Morville, Peter. 2014. *Intertwingled: Information Changes Everything*. Ann Arbor, MI: Semantic Studios.
Niño Moral, Dalila. 2008. "Un Estudio de Campo Realizado en Tres Hospitales de la Provincial de Alicante: Impresiones sobre el Papel del Intérprete." In *Investigación y*

Práctica en Traducción e Interpretación en los Servicios Públicos: Desafíos y Alianzas, edited by Carmen Valero-Garcés, Carmen Pena Díaz, and Raquel Lázaro Gutiérrez, 354–368. Madrid: Universidad de Alcalá.

Norman, Don A. 2004. *Emotional Design: Why We Love (or Hate) Everyday Things*. New York: Basic Books.

O'Brien, Elaine, Everlyn Nicodemus, Melissa Chiu, Benjamin Genocchio, Mary K. Coffey, and Roberto Tejada, eds. 2013. *Modern Art in Africa, Asia, and Latin America: An Introduction to Global Modernisms*. Hoboken, NJ: Blackwell Publishing.

O'Connor, Allison, Jeanne Batalova, and Jessica Bolter. 2019. "Central American Immigrants in the United States." Migration Policy Institute. https://www.migrationpolicy.org/article/central-american-immigrants-united-states-2017.

Quijano, Anibal. 2000. "Coloniality of Power, Eurocentrism, and Latin America." *Nepantla: Views from South* 1 (3): 533–580.

Redish, Ginny, and Carol Barnum. 2011. "Overlap, Influence, Intertwining: The Interplay of UX and Technical Communication." *Journal of Usability Studies* 6 (3): 90–101.

Ríos, Gabriela Raquel. 2015a. "Cultivating Indigenous Land-Based Literacies and Rhetorics." *Literacy in Composition Studies*, Special Issue: The New Activism 3 (1): 60–70.

Ríos, Gabriela Raquel. 2015b. "Performing Nahua Rhetorics for Civic Engagement." In *Survivance, Sovereignty, and Story: Teaching American Indian Rhetorics*, edited by Lisa King, Rose Gubele, and Joyce Rain Anderson, 79–95. Logan: Utah State University Press.

Rivera, Nora. 2020. "Chicanx Murals: Decolonizing Place and (Re)Writing the Terms of Composition." *College Composition and Communication* 72 (1): 118–149.

Rivera, Nora. 2022a. "Managing Indigenous Language Interpretation and Translation Services in the Public Sector." In *Indigenous Language Interpreters and Translators: Toward the Full Enactment of All Language Rights*, edited and translated by Abigail Castellanos García, Laura Gonzales, Cristina V. Kleinert, Tomás López Sarabia, Edith Matías Juan, Mónica Morales-Good, and Nora K. Rivera. Lexington, KY: Intermezzo.

Rivera, Nora. 2022b. "Understanding Agency through Testimonios: An Indigenous Approach to UX Research." *Technical Communication* 69 (4). https://doi.org/10.55177/tc986798.

Rivera, Nora, and Laura Gonzales. 2021. "Community Engagement in TPC Programs During Times of Crises: Embracing Chicana and Latina Feminist Practices." *Programmatic Perspectives* 12 (2): 39–65.

Rivera Cusicanqui, Silvia. 1987. "El Potencial Epistemológico y Teórico de la Historia Oral: De la Lógica Instrumental a la Descolonización de la Historia." *Temas Sociales* 11: 49–64.

Rivera Cusicanqui, Silvia. 2010. *Ch'ixinakax Utxiwa: Una Reflexión sobre Prácticas y Discursos Descolonizadores*. Buenos Aires: Tinta Limón.

Rodriguez, Eric, and Everardo J. Cuevas. 2017. "Problematizing Mestizaje." *Composition Studies* 45 (2): 230–233.

Rose, Emma J., Robert Racadio, Kalen Wong, Shally Nguyen, Jee Kim, and Abbie Zahler. 2017. "Community-Based User Experience: Evaluating the Usability of Health Insurance Information with Immigrant Patients." *IEEE Transactions on Professional Communication* 60 (2): 214–231.

Savage, Gerald, and Godwin Agboka. 2015. "Professional Communication, Social Justice, and the Global South." *Connexions: International Professional Communication Journal* 4 (1): 3–17.

Singhal, Arvind, and Lucía Durá. 2009. *Protecting Children from Exploitation and Trafficking: Using the Positive Deviance Approach in Uganda and Indonesia*. Fairfield, CT: Save the Children Federation.

Slack, Jennifer Daryl, David James Miller, and Jeffrey Doak. 1993. "The Technical Communicator as Author: Meaning, Power, Authority." *Journal of Business and Technical Communication* 7 (1): 12–36.

Sleasman, Brent C. 2015. "A Philosophy and Ethics of International Classroom Translation: Communicative Implications of Oral Mediation in Haiti." *Connexions: International Professional Communication Journal* 3 (2): 129–146.

Smith, Linda Tuhiwai. 2012. *Decolonizing Methodologies: Research and Indigenous Peoples*, 2nd ed. London: Zed.

Society for Technical Communication. 2023. "Defining Technical Communication." Society for Technical Communication. https://www.stc.org/about-stc/defining-technical-communication/.

St.Amant, Kirk. 2020. "Communicating about COVID-19: Practices for Today, Planning for Tomorrow." *Journal of Technical Writing and Communication* 50 (3): 211–223.

Stanford University d.school. 2020. "About." Stanford University d.school. Accessed September 15, 2020. https://dschool.stanford.edu/about.

Strowe, Anna. 2016. "Power and Conflict." In *Researching Translation and Interpreting*, edited by Claudia V. Angelelli and Brian James Baer, 118–130. New York: Routledge.

Sun, Huatong. 2006. "The Triumph of Users: Achieving Cultural Usability Goals with User Localization." *Technical Communication Quarterly* 15 (4): 457–481.

Suojanen, Tytti, Kaisa Koskinen, and Tiina Tuominen. 2015. "Usability as a Focus of Multiprofessional Collaboration: A Teaching Case Study on User-Centered Translation." *Connexions: International Professional Communication Journal* 3 (2): 147–166.

Tedlock, Dennis. 1996. *Popol Vuh: The Definitive Edition of The Mayan Book of the Dawn of Life and the Glories of Gods and Kings*. New York: Simon and Schuster.

Tham, Jason. 2022. "Past and Futures of Design Thinking: Implications for Technical Communication." *IEEE Transactions on Professional Communication* 65 (2): 261–279.

Tuck, Eve, and K. Wayne Yang. 2012. "Decolonization Is Not a Metaphor." *Decolonization: Indigeneity, Education and Society* 1 (1): 1–40.

Tyulenev, Sergey. 2016. "Agency and Role." In *Researching Translation and Interpreting*, edited by Claudia V. Angelelli and Brian James Baer, 17–31. New York: Routledge.

UN Commission on Human Rights. 1982. *Report of the Sub-Commission on Prevention of Discrimination and Protection of Minorities on Its 34th Session: Study of the Problem of Discrimination against Indigenous Populations*. New York: United Nations Economic and Social Council. https://www.un.org/development/desa/indigenouspeoples/publications/2014/09/martinez-cobo-study/.

United Nations. 2007. *United Nations Declaration on the Rights of Indigenous Peoples*. Accessed December 26, 2020. https://documents-dds-ny.un.org/doc/UNDOC/GEN/N06/512/07/PDF/N0651207.pdf?OpenElement.

Vespa, Jonathan, Lauren Medina, and David M. Armstrong. 2020. *Demographic Turning Points for the United States: Population Projections for 2020 to 2060*. Suitland, MD: United States Census Bureau. https://www.census.gov/library/publications/2020/demo/p25-1144.html.

Villanueva, Victor. 2016. "Forward." In *Decolonizing Rhetoric and Composition Studies: New Latinx Keywords for Theory and Pedagogy*, edited by Iris Ruiz and Raúl Sánchez, v–viii. London: Palgrave Macmillan.

Wadensjö, Cecilia. 2013. *Interpreting as Interaction*. New York: Routledge.

Walton, Rebecca. 2016. "Supporting Human Dignity and Human Rights: A Call to Adopt the First Principle of Human-Centered Design." *Journal of Technical Writing and Communication* 46 (4): 402–426.

Walton, Rebecca, Kristen Moore, and Natasha N. Jones. 2019. *Technical Communication after the Social Justice Turn: Building Coalitions for Action*. New York: Routledge.

Wible, Scott. 2020. "Using Design Thinking to Teach Creative Problem Solving in Writing Courses." *College Composition and Communication* 71 (3): 399–425.

Wilson, Shawn. 2008. *Research Is Ceremony: Indigenous Research Methods*. Nova Scotia: Fernwood Publishing.

Wynter, Sylvia. 2003. "Unsettling the Coloniality of Being/Power/Truth/Freedom: Towards the Human, after Man, Its Overrepresentation—An Argument." *CR: The New Centennial Review* 3 (3): 257–337.

Xiang, Zairong. 2016. "The (De)Coloniality of Conceptual Inequivalence: Reinterpreting *Ometeotl* through Nahua *Tlacuiloliztli*." In *Decolonial Approaches to Latin American Literatures and Cultures*, edited by Juan G. Ramos and Tara Daly, 39–55. London: Palgrave Macmillan.

Yajima, Yusaku, and Satoshi Toyosaki. 2015. "Bridging for a Critical Turn in Translation Studies: Power, Hegemony, and Empowerment." *Connexions: International Professional Communication Journal* 3 (2): 91–125.

Young, Iris Marion. 1990. "Five Faces of Oppression." In *Justice and the Politics of Difference*, 39–65. Princeton, NJ: Princeton University Press.

INDEX

Locators with an *f* indicate a figure. Locators with a *t* indicate a table.

academia, building alliances with Indigenous communities, 32*f*, 80, 81*t*, 82*t*, 130*f*
academic research: child language brokers, 43–44, 117; collaboration, 29; conduit model, 20–24; cultural usability framework, 26; decoloniality, 19; equity/inequity, 25, 132–136; orality, 12; social justice, 25; support, 6; technical and professional communication (TPC), 24–25. *See also* research methodology; scholarship; user experience (UX) research
accessibility, user experience (UX) research, 133*f*, 134
accreditation: government initiatives, 91, 114–116; Indigenous interpreters and translators, 87*t*–88*t*, 89*t*–90*t*. *See also* certification; professionalization
Achuar del Pastaza, 80, 81*t*, 82*t*, 83*t*, 117–118, 124, 136–139
acknowledgment of cultural differences, 145
activism: civic engagement, 68; interpreters and translators, 11, 103, 104*f*; Indigenous rights, 127*f*; injustices, 50*t*; language rights, 121
actor-network theory (ANT), 21–22, 68
advocacy: deeper-level needs, 51; design thinking process, 32*f*, 130*f*; dialogical interpreting practices, 67; diversity, 109; indigenous rights, 51*t*, 52, 62*t*, 87, 105, 106*f*, 130*f*; language rights, 12, 107–108, 121, 123; technical and professional communication (TPC), 108, 109; technology tools, 32*f*, 91, 130*f*; testimonios, 148
Africa, postcoloniality, 17
Afro Descendants, 153*n*4
agency, role of, Indigenous translation and interpreting practices, 4, 6, 21, 30
alliance-building with academia, 32*f*, 81*t*, 82*t*, 130*f*
Alonso, Icíar, 13*f*, 19, 21
alphabetic literacy, 55*t*
alphabetic writing, 19

ambiguity in interpreting events, 131–132, 150*f*
amoxcalli (libraries), 4
amoxtli (pictographic books), 3–4, 13
analog tools in interpretation and translation, 16, 46, 110
Angelelli, Claudia, 13*f*, 21–23, 43–44
antiracist pedagogies, 149–150
Anzaldúa, Gloria, 13*f*, 16, 17, 135, 154*n*1(chap. 1)
apprenticeships, Indigenous interpreters and translators, 139–142, 143*f*, 144
appropriation: academic research, 19; Indigenous languages, 138
Arias, Arturo, 21
audio-to-text transcriptions and translations, 39–40
autoethnographic research, 10–11
awareness/lack of awareness: cultural differences, 50*t*, 51*t*, 56, 60, 77, 98, 109; Indigenous matters, 73, 99*f*, 140; interpreters and translators, 75, 87*t*–88*t*, 130*f*; language rights, 32*f*, 45, 74*t*, 76*t*, 78*t*, 79*t*, 81*t*; law enforcement officers, 145. *See also* Indigenous culture; linguistic families
Aymara Indigenous practices, 13–14
Aztec culture, 3–4, 28, 61*t*

Baca, Damián, 13*f*, 19
Bakhtin, Mikhail, 22
Barnum, Carol, 25
Bhabha, Homi, 16, 19, 20
biases, 42, 94, 110
Biblioteca de Investigación Juan de Córdova, 29–30
bilingual programs, 17, 29, 48, 54
birth certificates, translation from Spanish, 124
blanquitud (whiteness), 94–95

"calling," Indigenous interpreters, 104*f*
Cardinal, Alison, 31
Castañeda, Antonia, 43
Castellanos García, Abigail, 132

166   INDEX

cathartic acts, 66, 68
Centro de Reinserción Social (Social Reintegration Center) (CERESO), 84, 122, 154n1(chap. 5)
Centro Profesional Indígena de Asesoría, Defensa y Traducción (CEPIADET), 29, 31, 35, 58, 133–136, 139–141
Certification for Healthcare Interpreters (CoreCHI), 89t, 90t, 116
certification: court interpreters, 79t, 113–115; government initiatives, 78t, 114–115; Indigenous languages, 89t, 90t, 120, 125–126; medical field, 116–117, registration, 87t, 88t; Spanish-language requirements, 115, 116. See also accreditation; professionalization
challenges, Indigenous interpreters and translators: research participants, 11, 63, 99f; user empathy maps, 36t, 46f, 47f, 48f, 49t, 51t, 52t–53t, 54t, 55t, 56t, 57t, 58t, 59t–60t, 61t, 62t, 99f; stereotypes, 16
Chávez Leyva, Yolanda, 18
Chi, Gaspar Antonio, 3
Chicanos/Chicanas/Chicanx, 17, 18, 154n1(chap. 2)
child language brokers, 11, 43–44, 52, 60, 61, 117
children's books, Indigenous language education, 65, 123, 140, 143–144
Chile, colonialism, 20
Chinanteco de San Pedro Yolox, 73, 74t, 75t, 111, 112t
citizenship in curricula, 150
civic engagement: activism, 68, 127; ideas, 70; intercultural translation projects, 127f; social change, 64, 66; testimonio maps, 38t, 74t–75t, 76t–77t, 78t–80t, 81t–82t, 84t, 86t, 88t, 90t, 92t
classrooms, user interface, 150f
co-constructing practice, 21–22
coatlicue (bilingualism), 17
Cocom, Juan, 3
code of ethics, 22, 23, 131–132, 147
Codex Borgia, 4
Colectivo Uantakua (*Traduciendo Juntas*), 142f, 143f
collective experiences: desahogo, 68; Indigenous interpreters and translators, 33; misalignment, 15; needs, 45; pain points, 70; research participants, 148; testimonios, 14, 64, 66, 68, 72f; struggle, 67–68; user empathy maps, 46, 47f, 72f; Western perspective, 149
collective initiatives, Indigenous cultures, 141–142

colonialism: decolonial theory, 154n1(chap. 1); effects on Indigenous culture, 3, 12, 15, 17, 20; ethnicities, 17; Eurocentric perspective, 19; marginalized communities, 102; Mestizes, 28–29; Otherness, 17; postcolonial theory, 154n1(chap. 1); systemic issues, 108; writing systems, 19
comic books, translation to Indigenous languages, 141
communal responsibility, Indigenous communities: assistance, 59, 60; connectedness, 106–107; deeper-level needs, 51; engagement, 15; Indigenous interpreters and translators, 45, 148; technical and professional communication (TPC), 96–97; *tequio*, 121
communication bridge, 103f, 104f
community-driven research, 15–16, 30
comuneras/comuneros (community members), 7
conduit model in scholarship, 20, 21, 22, 24
conferences, disruption of hierarchal roles, 153n3
Consejo Nacional de Normalización y Certificaión de Competencias Laborales (CONOCER), 113–114
consent of research participants, 35
content design. See user experience (UX) research
contributing to Indigenous communities, 105, 106f
Convention No. 169 (International Labor Organization), 122, 125–126
CoreCHI (Certification for Healthcare Interpreters), 89t, 90t, 116
Cortés, Hernán, 154n1(chap. 3)
Cosmic Race ("Raza Cósmica"), 18
cosmovision (worldview): Indigenous perspective, 118, 130, 131f, 138, 145; Mesoamerican cultures, 153n2; oral histories, 13–14; Western perspective, 44, 81t, 101, 130, 131f, 132
court interpreters: certification, 113–115; cultural differences, 70; distrust of legal system, 73; emotions, 69; international collaborations, 113; interpreting events, 69; oral desahogo, 69; protocols, 62t, 114; usability needs, 53; wages, 119–120. See also interpreters and translators
credibility, user experience (UX) research, 133f, 134–135
critical empathy, 10–11, 41–42
critical issues, Indigenous community, 6, 38t, 105, 106f, 111, 113f, 129, 130f

criticism of decolonial scholarship, 18
cross-cultural global research, 25
Cuevas, Everardo J., 18, 19
cultural context, user interface (UI), 30, 130, 145
cultural differences: awareness, 50*t*, 51*t*; colonialism, 17; deeper-level needs, 55; global migration, 31–32; imperialism, 98; Indigenous people, 49, 77, 79*t*; legal issues, 70; mediation, 144, 145; public officials, 50*t*; research participants, 20, 148; verticality, 131*f*
cultural records (*itolaca*), 3–4, 13
curricula, citizenship, 149–151

data coding, 11, 35, 37
death certificates, translation from Spanish, 124
decline of Indigenous language rights/use, 74*t*, 75, 76*t*
decolonial theories: frameworks, 15, 19–20; origin, 16; postcoloniality, 16, 154*n1*; research, 13*f*, 18; rhetoric and composition, 16; scholarship, 7–8, 10, 12, 17
deeper-level needs: community assistance, 47, 51, 59, 60; cultural awareness, 49, 55, 60; dialogue, 53; emotional preparation, 49; Indigenous rights, 52, 63; language preservation, 55; user empathy map, 45, 46*f*, 47*f*; validation for Indigenous professionals, 58
defining issues, 9, 30, 33, 37, 38*t*, 39
demographics of research participants, 95–96
Derrida, Jacques, 16
desahogo, 39, 64, 67–70, 105, 106*f*, 114, 134, 149
desahogarse, 10, 11, 66
design thinking process: Defining phase, 30, 31*f*, 32*f*, 39; Empathizing phase, 30, 31*f*, 32*f*, 33, 39, 42, 44–45; equity, 129–132; feelings, 102–103; Ideating phase, 30, 31*f*, 32*f*, 34, 39–40, 139–140, 141–144; intercultural, 136–139; language revitalization, 32*f*, 130*f*; localized, 39; Prototyping phase, 31*f*, 34–35; Re-designing phase, 32*f*, 34–35, 40, 97, 144–145; scholarship, 9–10; Stanford d.school, 31*f*; sticky notes, 45, 46*f*, 47*f*; Synthesizing phase, 32*f*, 40, 121–122; testimonios, 32*f*, 33, 64–65; training, 32*f*, 45, 120–122; user experience (UX) research, 30, 132–136, 148
desirability, user experience (UX) research, 133*f*, 134

DeWitt, Richard, 153*n2*
dialogue: deeper-level needs, 53; Indigenous practice, 65–67, 105, 106*f*; living, 110; promoting, 111; research methods, 149; synthesized, 15; untranslatable terms, 67
Díaz, Porfirio, 154*n3*
digital technology, 16, 134–135, 139, 149
disabilities, people with, 134
discrepancies, interpreting and translation, 21
discrimination, Indigenous people, 6, 48*f*, 52, 75, 76*t*, 83, 87*t*–88*t*, 111, 122–124, 126, 129, 130*f*, 147
disruption of dominant narratives, 8, 9, 25, 148, 153*n3*
distrust of court systems, 73
diverging/converging sessions, 30–31, 34
diversity: advocacy, 109; curricula, 150–151; design thinking process, 32; Latin Americans, 153*n4*; superdiversity, 31–32, 148
Doak, Jeffrey, 24, 67
dominant narratives, 25
Doña Marina, 52*t*, 154*n1*(chap. 3)
double-edged service, 104*f*

Echeverría, Bolívar, 94–95
education. *See* training
effectiveness, technical and professional communication, 25
efficiency, interpretation and translation services, 77
emerging technologies, 144
emotional maturity/emotional toll, 21, 49, 50*t*, 69, 99*f*, 101, 109, 131–132, 134
Empathizing phase (design thinking process): critical empathy, 41–42; design thinking process, 31*f*, 32*f*; Empathy maps, 9, 10–11, 44–45, 47*f*, 63; Indigenous interpreters and translators, 39; interviews, 30, 32*f*, 33, 42; user experience (UX) research, 42
employment conditions/opportunities, 11, 140, 144–145
empowerment, Indigenous language speakers, 77*t*
English language, 96*f*, 115, 116
equity, user experience (UX) research, 6, 30, 110, 129–132, 133*f*, 134–136, 148, 149
Escuela Nacional de Estudios Superiores (ENES) Campus Morelia, 141, 142*f*, 143, 144
ethical responsibilities, user experience (UX) research, 24, 44, 108, 117, 134–135

ethnicities, 17, 149–150
Eurocentric perspective, 19, 145
European languages, 20, 100–101
Evaristo Villegas, José Luis, 31
exclusion of Indigenous cultures, 98

family interpreters, 44, 51
Federación de la Nacionalidad del Pueblo de Achuar del Perú (FENAP), 136, 138
feelings, Indigenous interpreters and translators, 11, 47*f*, 102, 103*f*
Felipillo, 52*t*, 154*n1*(chap. 3)
fields of work, interpreting and translation, 36*t*, 38, 46*f*, 48*f*, 49*t*, 51*t*, 52*t*–53*t*, 54*t*, 55*t*, 56*t*, 57*t*, 58*t*, 59*t*–60*t*, 61*t*, 62*t*, 71*f*, 87*t*, 96*f*, 112*t*
findability, 133*f*
Fogg, B. J., 134
foreign languages interpretation and translation services, 136
formation of worldviews, 22–23
Fuentes Gómez, Luis Arturo, 31
funding for Indigenous interpreters and translators, 45, 77, 120–122, 136, 145

gender, research participants, 96*f*, 112*t*, 150, 153*n4*
Gentile, Paola, 105
geographical inequalities, 153*n5*
Giddens, Anthony, 21–22
global migration, 31–32
Global South, 25, 153*n5*
glossaries, 39, 40*f*, 80, 82*t*, 137
Gonzales, Laura, 13*f*, 24, 26, 153*n6*
government documents, 55, 80, 82*t*, 124, 136–139
government initiatives, 20, 21, 22, 24, 52, 54, 83, 91, 114–115, 117–118
government policies, 68, 78*t*, 79*t*, 86*t*, 87*t*–88*t*, 111, 121–122, 124, 125, 126, 130*f*

Haas, Angela M., 13*f*, 16, 94
healthcare. *See* medical field
hegemonic epistemologies, 17
hierarchal roles, disruption of, 153*n3*
Hierholzer, Kirstin, 153*n6*
high-school level requirements, 115
historical context, Indigenous languages, 145, 149
history (*itolaca*), 3–4, 13
honeycomb, user experience (UX), 133*f*
Hormiguera Editorial, 141
huehuehtlahtolli (wise dialogues), 14, 15, 19
humana/humano, 154*n2*
human rights organizations, 145
human-centered design, 25

I, Rigoberta Menchú: An Indian Woman in Guatemala, 15
Ideating phase (design thinking process): apprenticeships, 31*f*, 32*f*, 141–144; diverging/converging sessions, 30–31, 34; intercultural translation projects, 136–139; past-tested ideas, 38*t*, 39–40, 70, 72*f*; presentations, 32*f*; thinking, 31; user experience (UX) research, 134
identifying themes with sticky notes, 70, 71*f*, 72*f*
identity: in-betweenness, 17; Indigenous people, 106–107; Mestizes, 28–29; post-coloniality, 16; testimonios, 38
imbalances in institutional verticality, 131–132
immediate context, Indigenous scholars, 7–8
impartiality/partiality of interpreters and translators, 8, 20, 22, 23, 67, 110, 144, 147
imperialism, effects on Indigenous culture, 12, 98
implications, testimonio maps, 38*t*, 71*f*, 72*f*
in-betweenness, mixed identities, 17
INALI (Instituto Nacional de Lenguas Indígenas), 100, 113–114, 153*n1*
incarceration of Indigenous people, 59*t*–60*t*, 75*t*, 98, 139, 140
Inclán Solís, Daniel, 20
inclusion, 8, 132, 137, 139
Indigenous culture: awareness, 55, 56, 59, 60, 73, 74*t*, 76*t*, 78*t*, 79*t*, 81*t*, 99*f*, 101, 144; collective initiatives, 141–142; colonialism, 7–8, 10, 12, 13*f*, 15, 17, 108; communication bridge, 104*f*; connectedness, 45, 105–107; cosmovision, 130, 131*f*; disenfranchisement, 147; exclusion, 98; history, 18, 149; invisibility/visibility, 21–22, 85, 104; linguistic needs, 97–98; local authorities, 69; needs, 97*f*; public institutions, 138; testimonios, 112*t*
Indigenous interpreters and translators. *See* interpreters and translators
Indigenous languages: appropriation, 138; glossaries, 82*t*; marginalization, 76*t*, 83, 102; monolingual speakers, 27; oral histories, 13–14, 19, 29; political context, 149; protocols, 82*t*; revitalization, 77, 130*f*, 140, 142–143; translating, 4, 20, 21, 65–66, 141, 144; value, 100–101, 136; visibility, 107–108. *See also* linguistic families
Indigenous Peoples: advocacy, 62*t*, 130*f*; communicating with Spanish speakers, 19–20; community-oriented roles, 148; cultural awareness, 49; digital technical

Index   169

skills, 139; discrimination, 48*f*, 52, 76*t*, 97*f*, 111, 123, 129; government policies, 52, 78*t*, 79*t*, 86*t*, 87*t*, 88*t*, 117–118, 122, 139; ignorance, 73, 99; incarceration, 59*t*–60*t*, 75*t*, 98, 139, 140; legal issues, 57*t*, 73, 62*t*, 74*t*, 104; migrant communities, 50–51, 120, 126; multilingualism, 75–77; Otherness, 17; public sector, 62*t*; reintegration centers, 74*t*; school districts, 120; self-identity, 15, 105, 106*f*, 107, 154*n1*(chap. 2); stereotypes, 16
Indigenous perspective, 39, 132, 144, 145, 149
Indigenous rights: advocacy, 52, 87, 105, 106*f*, 109, 127*f*; awareness, 32*f*, 140; language rights, 12–13, 32*f*, 51*t*, 52*t*–53*t*, 77*t*, 83*t*, 87, 124–126, 132, 138; marginalization, 108; United Nations Declaration on the Rights of Indigenous Peoples, 135; validation, 58
Indigenous scholars: alliance with academia, 29, 80, 81*t*, 130; benefits, 107; immediate context, 7, 8; perspective, 80; positionality, 41–42; professionalization needs, 111, 124, 146; user experience (UX), 133–134, 146
individual testimonio maps, 70, 71*f*
inequalities, 25, 122, 153*n5*
Inghilleri, Moira, 22, 23, 67, 110
injustices, 50*t*, 97*f*
Institution Review Board (IRB), 35
institutional verticality, 110, 130, 131*f*, 132
Instituto Internacional de Derecho y Sociedad (IIDS), 136, 138
Instituto Nacional de Lenguas Indígenas (INALI), 31, 100, 113–114, 153*n1*
intercultural translation projects, 10, 80, 124, 127*f*, 136–139
international collaborations, 87, 113, 135
International Labor Organization, 122, 125–126
International Unconference for Indigenous Interpreters and Translators, 5–6, 9, 29, 30, 33, 35, 39, 44
interpersonal skills, 22
interpreters and translators: activism, 11, 127; apprenticeships, 139–142, 143*f*, 144; awareness, 75, 130*f*; certification, 79*t*, 91, 120, 125; child language brokers, 43–44, 117; code of ethics, 22; collective experiences, 33; communication bridge, 103*f*, 104*f*; cosmovision, 22–23, 131*f*, 138; court interpreters, 73; cultural issues, 20, 79*t*; deeper-level needs, 6, 45, 106*f*, 113*f*; discrimination, 75, 83, 87*t*–88*t*, 111, 129; diversity, 31; emotions, 50*t*, 99*f*; employment, 144–145; government policies, 83, 87*t*–88*t*, 121–122, 136; partiality, 22, 23, 108, 144; inclusion, 137; legal sector, 79*t*, 140–141; mediator role, 5, 23, 131–132, 135, 145; motivation, 43, 44, 63, 96, 97*f*, 98; practical, 79*t*; professionalization, 21, 26, 32*f*, 76*t*, 77, 89, 112–113, 124–126; recognition, 48*f*, 49*t*, 52*t*–53*t*, 55, 60; school districts, 91; self-perception, 103, 104*f*, 105; training, 22, 23, 45, 49*t*, 54*t*, 75, 99*f*, 109, 115, 121; usability needs, 43, 45; value, 55*t*, 57*t*; wages, 45, 68, 77, 85, 87*t*, 101, 102, 118–119
interpreting and translating: community needs, 106–107; conduit model, 20, 21, 22, 24; contexts, 144, 145; decoloniality, 20; desahogo, 15, 105, 106*f*; discrepancies, 21; efficiency, 77; Eurocentric perspective, 145; government documents, 54–55; Indigenous perspective, 26, 144, 145; institutional verticality, 132; localization, 26; neutrality, 20, 22; oral languages, 16; politicization, 20; postcoloniality, 20; software programs, 137–138; standardizing, 20; technology, 16, 137–138; written languages, 16
interpreting events: ambiguity, 131–132; biases, 110; contexts, 30; court interpreters, 69, 73; desahogo, 67–70, 105, 106*f*, 110, 114; dialogue, 66–67, 105, 106*f*; emotional toll/training, 49, 101, 109, 114, 131–132; equity, 110; institutional verticality, 132; inter-sender conflict, 103; legal sector, 102, 140–141; medical sector, 116–117; neutrality, 67, 110; power dynamics, 67; rhetorical mediators, 110; testimonios, 38*t*, 66; vertical user interface, 130, 131*f*; video translation, 65–66; Western cosmovision, 101, 114, 130, 131*f*
inter-sender conflict, 103
interview participants. *See* research participants
introductions, protocols, 37
invisibility/visibility of Indigenous communities, 21–22, 85
issues in collective testimonio maps, 39, 40, 71*f*, 72*f*
Itchuaqiyaq, Cana Uluak, 107
itolaca (culture and history), 3–4, 13

Jaqaru linguistic family, 54, 55*t*, 96*f*, 101, 104, 107
jargon, 94
Jones, Natasha N., 13*f*, 24–25, 95, 100, 110

170   INDEX

K'iche' people, 4, 110
Kleinert, Cristina, 13f, 19, 21, 109
Koskinen, Kaisa, 25

La Malinche, 52t, 154n1(chap. 3)
Ladina/Ladino. *See* Mestiza/Mestizo
Landa, Diego de (Fray), 3
languages: barriers, 50, 51t, 52t–53t, 107; language rights laws, 32f, 87t–88t, 123, 145; localization, 139, 153n6; origin, 96f, 112t; politics of, 19; revitalization projects, 15, 32f, 55, 76t, 127f, 130f; testimonio maps, 38, 71f; user empathy maps, 36t, 46f, 48f, 49t, 54t, 57t, 58t, 61t, 62t, 87t; variants, 55t, 56t, 59t–60t, 89, 100
Latifundios, 154n3
Latin America, 17, 27, 29, 153n4
Latinasians, 153n4
Latinxs, 17
Latour, Bruno, 21–22, 68
law enforcement officers, 145
Layme Yépez, Hernán, 31
legal issues: child language brokers, 43–44, 52, 61, 117; court protocols, 63; desahogo, 69–70; employment opportunities, 139–141; glossary of terms, 39, 40f; Indigenous people, 57t, 73, 74t, 104, 140, 145; institutional verticality, 131f; interpreters and translators, 60, 79t, 83, 102, 139–141; language discrimination, 107, 111, 123; linguistic rights, lack of, 125–126; power dynamics, 131; professionalization, 50, 111, 118; technical and professional communications (TPC), 140–141; terminology, 58t; untranslatable terms, 59
lenguas indígenas/lenguas originarias, 7
León-Portilla, Miguel, 13f, 14
libraries (*amoxcalli*), 4
lingua franca, 4, 19–20
linguistic activism, 108, 121, 143–144
linguistic families: Achuar, 80, 81t, 82t, 83t, 117–118, 124, 136–139; Aymara, 13–14; Chinanteco de San Pedro Yolox, 73, 74t, 75t, 111, 112t; Jaqaru, 54, 55t, 96f, 101, 104, 107; Mazateco, 58t, 96f; Mixteco, 1, 46f, 49, 50t, 51t, 57t, 71f, 77, 78t, 79t, 80t, 96f, 97; Quechua, 44, 52t, 61, 62t, 88t, 96f, 102, 123; Tarahumara, 27, 28, 83, 84t, 111, 112t, 154n1(chap. 5), 155n1; Tarasco, 65, 128, 141, 142f, 143f, 144, 155n1; Triqui, 60, 61t, 86t, 96f, 97, 126; Tzeltal, 56t, 85f, 86t, 95, 96f, 104, 111, 112t, 126; Yuto-Nahua, 3, 4, 13, 20, 46f, 48f, 95, 96f, 106, 108, 153n1;

Zapoteco, 53, 54t, 59t, 71f, 75, 76t, 77t, 95, 96f, 98, 111, 123. *See also* Indigenous languages
linguistic rights, 49, 63, 87, 97–98, 124, 139–141
linguistic sovereignty, 12–13, 15, 53t, 87, 108, 125–126, 130f, 132, 135, 136–139, 145
linguistic standards, 144
linguistic verticality, 131f
literacy, 19, 55t, 143
lived experiences, 66
living dialogues, 110
local laws, Indigenous rights, 69, 87, 121–122, 135
localization, research methods, 8–9, 26, 39, 153n6
loose professionalization systems, 112–113
Lugones, María, 98
Luhmann, Niklas, 21–22

Malintzin, 52t, 154n1(chap. 3)
marginalization, Indigenous languages, 8, 24, 75, 76t, 83, 102, 108, 123, 129–130
marriage certificates, 124
Martínez Cobo Study, 7
Masehuatl Tajtol, 153n1
Matheson, Breeanne, 107
Matías Juan, Edith, 130, 132
Mayan writing system, 3, 100
Mazateco, 58t, 96f
meaning constructs, 4
mediator, as role for Indigenous translators and interpreters, 5, 23, 131–132, 135
medical field, 6, 50, 60, 69, 107, 111, 116–117, 118, 119
Memsource, 83t, 137–138
Menchú, Rigoberta, 13f, 15, 106–107
mentoring projects, 30, 127f
Mesoamerican cultures, 3–4, 153n2
Mestiza/Mestizo: colonialism, effects of, 28–29; decolonial theories, 12; in-betweenness, 17; Indigenous scholarship, 41–42; positionality, 18–19; self-identity, 28–29, 153n4, 154n1(chap. 2); Western traditions, 27
methodologies, 144
Mexcatl, 153n1
Mexica. *See* Aztec culture
Mexican Americans, 17, 18, 28, 154n1(chap. 2)
Mexicano, 153n1
Mexico: Chicanos/Chicanas/Chicanx, 17, 18, 154n1(chap. 2); colonialism, 19–20; court interpreters, 113–114; government initiatives, 121, 128; *itolaca*, 13; language

variants, 100; *lenguas indígenas*, 7; linguist rights, 5, 59, 125–126, 139–141, 153*n1*; Mexican Revolution, 154*n3*; nationalism, 154*n3*; professionalization, 89; translation projects, 136; user experience (UX) research, 134
Michoacán, 32, 91, 136, 141–144
Mignolo, Walter, 13*f*, 19, 154*n1*
migrant Indigenous communities, 43, 50–51, 56, 85, 86*t*, 91, 92*f*, 117, 120, 126
Miller, David James, 24, 67
misalignment, collective exercises, 15
misappropriation, Indigenous knowledge, 18
mistranslations of Spanish texts, 18
mixed identities, in-betweenness, 17
mixed race. *See* Mestiza/Mestizo
Mixe language, 151
Mixteco linguistic family, 1, 46*f*, 50*t*, 51*t*, 57*t*, 71*f*, 78*t*, 79*t*, 80*t*, 96*f*
Mixteco Indígena Community Organizing Project (MICOP), 29
monocultural/monolingual concepts, 17, 27, 29
Moore, Kristen R., 95, 100, 110
Moraga, Cherríe, 16, 135
Morales-Good, Mónica, 3, 41, 121
Morville, Peter, 129, 133*f*, 134–135
motivation: Indigenous interpreters and translators, 11, 43, 44, 47*f*, 96, 97*f*, 98; user empathy maps, 36*t*, 46*f*, 48*f*, 49*t*, 51*t*, 52*t*–53*t*, 54*t*, 55*t*, 56*t*, 57*t*, 58*t*, 59*t*–60*t*, 61*t*, 62*t*
multiculturalism, 7, 55*t*, 145, 149–150
multidimensional cultures, 99
multilingualism, 7, 24, 55*t*, 75–77, 83, 85, 145, 149–150
multimodality, 24
multinational master list, 144

Nahua/Nahuatl/Nauta. *See* Yuto-Nahua linguistic family
narratives. *See* testimonios
nationalism, 154*n3*
national laws, 87, 135
Native Americans, 7, 29
Native languages, 52*t*–53*t*
Native Nations. *See* Indigenous Peoples
Native Peoples, 52*t*–53*t*
needs. *See* deeper-level needs; usability needs
Neltiliztli (fixed roots of oral history), 14
Nepantla (in-betweenness), 17
neutrality. *See* impartiality/partiality
Niño Moral, Dalila, 21, 23, 101
non-Indigenous cosmovision, 130, 131*f*

non-Indigenous decolonial theories, 7–8, 10, 12, 13*f*, 17, 37
non-Indigenous research participants, 19, 70
non-Western perspectives, 26
nonprofit organizations, 135

Oaxaca, 75–76, 77–80, 139–141
oppression, political, 66
oral histories, 12–14, 16, 19, 24, 29, 49, 94
origins, decoloniality, 16–17
Otherness, 16, 17
outcomes, testimonio maps, 38*t*, 75*t*, 77*t*, 82*t*–83*t*, 84*t*, 86*t*, 88*t*, 90*t*, 92*t*

pain points, testimonio maps, 38*t*, 70, 71*f*, 74*t*, 76*t*, 78*t*–79*t*, 81*t*, 84*t*, 85*f*–86*t*, 87*t*–88*t*, 89*t*–90*t*, 91*f*–92*f*
participants. *See* research participants
participatory research methods, 8–9, 15–16
Payás, Gertrudis, 13*f*, 19, 21
payment. *See* wages
perceptions of Indigenous interpreters and translators, 103
Pérez Holguín, Guadalupe, 131
persona creation, 30, 33
personal narratives, 15, 64, 67–68, 148
persuasive computers, 134–135
Peru, 7, 122, 124–126, 136–139
phonograms, 3
pictographic books (*amoxtli*), 3–4, 13
place of origin, user empathy maps, 36*t*, 38*t*, 46*f*, 48*f*, 49*t*, 51*t*, 52*t*–53*t*, 54*t*, 55*t*, 56*t*, 57*t*, 58*t*, 59*t*–60*t*, 61*t*, 62*t*, 71*f*, 74*t*, 76*t*, 78*t*, 81*t*, 87*t*, 89*t*, 91*f*–92*f*, 96*f*, 112*t*
plain language protocols, 137
political context, 149
politicization of interpreting and translation, 19, 20, 24
Popol Vuh, 4
positionality, 18–19, 24–25, 27–29, 41–42
Postclassic period, 3
postcoloniality, 16–17, 19, 20, 154*n1*
power dynamics: academic research, 25; Indigenous interpreters and translators, 131–132; Indigenous-Mestize interactions, 28–29; interpreting events, 67; legal sector, 131; privilege, 100; Rhetorical Negotiation Web (RNW), 131*f*; systemic issues, 129–130; user experience (UX) research, 135
practical training, usability needs, 59*t*, 60, 79*t*, 113–114
practicum, certification, 114–116
prejudices, 42
priorities, technical communication, 118

privilege, 27–29, 100, 110
Process phase (design thinking process), 31*f*, 32*f*
products, contents, processes, user experience (UX) research, 9, 130, 133, 134, 148
professional advancement, 54*t*, 97*f*
professional contexts, 22, 30, 144
professionalization, 109, 111, 124; code of ethics, 22, 23; development, 120–121; follow-up, 83, 85; government initiatives, 78*t*, 111, 117–118; interpreters and translators, 26, 32*f*, 40, 45, 75, 76*t*, 77, 81*t*, 91, 108, 130*f*, 133–134, 137; loose systems, 112–113, 124; medical field, 111; scholars, 146; standardizing, 113; updated training, 109; wages, 83, 84*t*, 92*f*, 108; workshops, 80*t*. *See also* accreditation; certification
protection, Indigenous language rights, 83*t*, 87*t*
protocols, 36–37, 63, 82*t*
Prototyping phase (design thinking process), 9, 31*f*, 34–35, 148
pseudonyms of research participants, 35, 48
public officials, 50*t*
public sector: awareness of Indigenous language rights, 32*f*; building trust, 138, 147; employment conditions/opportunities, 144–145; Indigenous interpreters and translators, 32*f*, 62*t*, 103, 105
pueblos indígenas/pueblos originarios, 7
P'urhepecha linguistic family, 65, 128, 141, 142*f*, 143*f*, 144, 155*n1*

qualitative research methods, 35
Quechua linguistic family, 44, 52*t*, 61, 62*t*, 88*t*, 96*f*, 102, 123
Quijano, Anibal, 13*f*, 17, 108

racial identity, 17, 28–29
"Raza Cósmica" (Cosmic Race), 18, 154*n3*
Re-designing phase (design thinking process), 32*f*, 34–35, 40, 97, 144–145, 150–151
recognition of Indigenous interpreting and translation, 48*f*, 49*t*, 52*t*–53*t*, 55, 99*f*, 101, 105, 106*f*
Reddy, Michael, 20
Redish, Ginny, 25
Registro Nacional de Identificación y Estado Civil (RENIEC), 138
reintegration centers, 74*t*, 125, 140, 154*n1*(chap. 5)
research methodology: autoethnographic, 10; child Indigenous translators, 43–44, 117; community-driven, 15; decolonial, 15; desahogo, 149; digital video conferencing tools, 149; empathy maps, 11; interviews, 35, 45; localization, 8–9; oral histories, 12–14; participatory action, 8–9, 15–16; sticky notes, 45, 46*f*, 47*f*, 70, 71*f*, 72*f*; technical and professional communication (TPC), 13*f*; testimonios, 14, 15, 35, 148–149; translating and interpreting studies (TPS), 13*f*; user experience (UX), 6, 153*n6*. *See also* academic research; scholarship; user experience (UX) research
research participants: career fields, 96*f*, 112*t*; challenges, 11, 63, 99*f*; civic engagement, 38; consent, 35; demographics, 95–96; feelings, 11, 103*f*; interview questions, 32*f*, 33, 35, 36–37, 39–40, 44–45; languages, 38, 96*f*, 112*t*; motivations, 11, 63, 97*f*; narratives, 148; non-Indigenous, 70; pain points, 38, 111; place of origin, 96*f*, 112*t*; protocol introductions, 37; pseudonyms, 35, 48; self-identity, 38; self-perceptions, 11, 104*f*; shadowing, 30, 42; superdiversity, 148; usability needs, 37. *See also* user empathy maps
resistance as a metaphor, 17
respect, Indigenous interpreters and translators, 99*f*
responsibility, sense of, Indigenous communities, 45
revitalization of Indigenous languages, 32*f*, 77, 127*f*, 130*f*, 140, 142–143
rhetoric and composition, 16, 17, 149–150
rhetorical negotiation web (RNW), 110, 131*f*, 132, 150*f*
rhetorical sovereignty, 15, 26
rhetorical traditions, 4
Ríos, Gabriela Raquel, 18
rituals, recording of, 3
Rivera Cusicanqui, Silvia, 13*f*, 14, 15, 17, 20, 67
Rodriguez, Eric, 18, 19
role of Indigenous interpreters and translators, 6, 30, 31, 33, 103–105
rural/urban Indigenous interpreters and translators, 119

salary. *See* wages
scholarship: benefit for Indigenous Peoples, 107; design thinking process, 10; Indigenous histories, 18; intercultural, 10; questioning, 42; technical and professional communication (TPC), 12; translation and interpreting studies

Index 173

(TIS), 12. *See also* academic research; research methodologies; user experience (UX) research
school districts (US), 91, 120–121, 126
selection process, 21–22
self-identity, Indigenous people, 11, 15, 38, 103, 104*f*, 105–107, 154*n1*(chap. 2)
semistructured interviews, 35–37
sexuality, curricula, 150
shadowing, 30, 42
sign language interpretation, 8
simplifying government documents, 80
Slack, Jennifer Daryl, 24, 67
Sleasman, Brent C., 24
Smith, Linda Tuhiwai, 12, 13*f*, 66, 106
social change, 64, 66, 68
social contexts, 145
social justice advocacy, 24–25, 94, 109
social media, 141, 142*f*
societal imbalances, 100
sociology framework, 21–22
sociopolitical contexts, 30, 130, 131*f*, 144
software programs, 137–138
South Asia, 17
Spanish language, 4, 18–20, 40*f*, 96*f*, 115, 116, 141
Spelman, Elizabeth V., 98
Spivak, Gayatri, 16
standardizing professional systems, 20, 113
Stanford d.school, 9, 30, 31*f*, 37, 42
statistics, Indigenous language awareness, 74*t*
stereotypes, Indigenous communities, 16
sticky notes, research method, 45, 46*f*, 47*f*, 70, 71*f*, 72*f*
story mapping, 30
storytelling. *See* testimonio maps
strategies, design thinking process, 34–35, 40, 154*n2*(chap. 2)
stress during interpreting events, 49, 114
Strowe, Anna, 23
structural oppression, 24
style guides, 80, 82*t*, 137
Sun, Huatong, 13*f*, 26
Suojanen, Tytti, 25
superdiversity, 31–32, 148
support, Indigenous language variants, 50*t*, 59
Synthesizing phase (design thinking process): dialogues, 15; feelings, 102–103; government policies, 121–122; motivations, 96–98; needs and issues, 32*f*, 40; sticky notes, 70, 71*f*, 72*f*
systemic issues, coloniality, 41, 108, 129–130
systemic oppression, 110

talking circles, 15
Tarahumara linguistic family, 27, 28, 83, 84*t*, 111, 112*t*, 154*n1*(chap. 5), 155*n1*
Tarasco linguistic family, 65, 128, 141, 142*f*, 143*f*, 144, 155*n1*
technical and professional communication (TPC): advocacy, 108, 109; alliances with Indigenous organizations, 145; antenarratives, 25; antiracist pedagogies, 149–150; apprenticeships, 143–144; community-based projects, 96–97; dialogues, 111; effectiveness, 25; equity, 129–132; ethical responsibilities, 24; Global South perspectives, 25; government documents, 139; inclusion, 132; interpreting events, 140–141; multiculturalism, 24, 145; oral communication, 94; priorities, 118; research, 8–9, 12, 13*f*; translation and interpreting (T&I), 5, 6, 144; user experience (UX) research, 25–26; Western perspective, 95
technical documents, designing/redesigning, 97
technology tools for advocacy, 32*f*, 91, 120, 127, 130*f*, 131*f*
tequio, 121
terminology, untranslatable, 54*t*, 58*t*
test, design thinking process, 31*f*
testimonio maps, 38*t*, 70, 71*f*, 74*t*, 75*t*, 76*t*, 80*t*, 82*t*, 84*t*, 86*t*, 88*t*, 90*t*, 92*t*, 127; collective experience, 14, 64, 66–68, 72*f*, 148; data coding, 37; field of work, 112*t*; outcomes, 77*t*, 83*t*; pain points, 78*t*–79*t*, 81*t*, 85*f*, 87*t*, 90*t*, 91*f*–92*f*; themes, 73
testimonios: audio-to-text transcriptions, 39–40; decolonial research, 15; desahogo, 10, 64, 66–70; design thinking process, 32*f*, 33, 64–66; dialogue, 65–67, 111; ideas tested, 38*t*; issues identified, 38*t*; metatestimonio, 68; pain points, 9, 37, 38*t*, 111; personal narratives, 14, 67–68, 148; perspectives, 37, 112*t*, 149; place of origin, 112*t*; prompts, 37; research methods, 15, 35, 148–149; rhetorical sovereignty, 15; Zapoteco linguistic family, 75–77
testing role-playing, 9, 31
themes in testimonio maps, 73
theoretical frameworks, 15–16, 18
theory of action, 21–22
thinking, ideating, 31
TIS. *See* translation and interpretation studies (TIS)
Tla'tol, 153*n1*
Tlacuilolitztli, 3–4

Tlacuilos (Aztec writers), 4
Tlamatinime (Nahua philosophers), 14
Toyosaki, Satoshi, 24
TPC. *See* technical and professional communication (TPC)
traditional context, user interface (UI), 130
Traduciendo Juntas (Translating Together), 141, 142*f*, 143*f*, 144
training: children, 123; emotional, 109; funding, 45, 120–122, 145; Indigenous interpreters and translators, 22, 23, 32*f*, 75, 99*f*, 101, 105, 106*f*, 115, 124; language barriers, 107; professionalization, 50, 111, 117, 118; updated, 109; usability needs, 59*t*, 60; user interface (UI) models, 149–150; wages, 119. *See also* accreditation; certification; professionalization
transcription and translation, 39–40, 45
transformation of worldviews, 22–23
translating children's books, 65
translation and interpretation studies (TIS): co-constructing practice, 21–22; code of ethics, 147; decolonial frameworks, 19–20; equity, 149; inclusion, 132, 139; research literature, 13*f*; scholarship, 12, 80; sociology framework, 21–22; support, 6; usability methods, 25–26; user experience (UX), 9; Western practices, 6
translation and interpreting (T&I) practices, historical, 8, 19–20
translation projects, 8, 10, 39–40, 124, 139, 141–144
translators. *See* interpreters and translators
travel expenses for court reporters, 119
Triqui linguistic family, 60, 61*t*, 86*t*, 96*f*, 97, 126
Tuominen, Tiina, 25
Tzeltal linguistic family, 56*t*, 85*f*, 86*t*, 95, 96*f*, 104, 111, 112*t*, 126

unconference format, 30, 153*n*3
underrepresented groups, 148, 150*f*
United Nations (UN) reports, 7, 135
United States, 5, 8, 17, 27, 30, 31, 39, 41, 75, 76*t*, 80, 89, 90, 92*f*, 96*f*, 100, 112*t*, 115, 120, 154*n1*(chap. 2)
Universidad Autónoma Benito Juárez de Oaxaca (UABJO), 139–141
Universidad Nacional Autónoma de México (UNAM), 141, 142*f*, 143–144
Universidad Peruana de Ciencias Aplicadas (UPC), 29, 136
university classrooms, 150*f*

untranslatable terms, 53, 54*t*, 58*t*, 59, 67
updated training, 105, 106*f*, 109
urban/rural Indigenous interpreters and translators, 119
usability needs: content, 153*n*6; court interpreters, 53; cultural, 26; culture, knowledge of, 51; Indigenous language variants, 55, 56; linguistic preparation, 49; practical training, 59*t*, 60; recognition, 52, 55; research participants, 37; technical and professional communication (TPC), 25–26; translation pedagogy, 25–26; untranslatable terms, 59; updated protocols, 63; user empathy map, 46*f*, 47*f*; user experience (UX) research, 44–45, 61*t*, 133*f*, 134; wages, 58, 111
user empathy maps, 36*t*, 46*f*, 47*f*, 48*f*, 49*t*, 51*t*, 52*t*–53*t*, 54*t*, 55*t*, 56*t*, 57*t*, 58*t*, 59*t*–60*t*, 61*t*, 62*t*, 99*f*; collective maps, 72*f*; data coding, 35; deeper-level need; field, 87*t*; interview participants, 71*f*; languages, 87*t*; motivation, 97*f*; participant interviews, 44, 63; sticky notes, 45. *See also* research participants
user experience (UX) research: accessibility, 133*f*; advocacy, 108; children's books and videos, 143–144; collective experiences, 64; Empathize phase, 42; equity, 30, 132, 148; ethical responsibility, 129–130; human-centered design, 25; inclusion, 139; Indigenous scholarship, 6, 9, 146, 147; localization, 26, 153*n*6; multicultural approaches, 145; products, contents, processes, 9, 130, 133, 134, 148; research methodology, 153*n*6; technology, 25, 118, 134–135; testimonios, 9–10, 15; usability, 6, 44–45. *See also* academic research; research methodologies; scholarship
user interface (UI), 130, 131*f*, 149, 150*f*
user-centered design (UCD), 153*n*6
user-centered translation (UCT) approach, 26

validation of Indigenous professionals, 58
value: Indigenous languages, 55*t*, 57*t*, 100–101, 105, 106*f*, 136; user experience (UX) research, 133*f*, 135
variants. *See* linguistic families
Vasconcelos, José, 18, 154*n*2
vertical user interface (UI), 110, 130, 131*f*, 132
video technology, 65–66, 120, 143–144
visibility, Indigenous culture, 31–32, 60, 63, 104, 105, 106*f*, 107–108

Wadensjö, Cecilia, 13*f*, 22, 67
wages, 57*t*, 58, 68, 77, 79*t*, 83, 85, 87*t*, 101, 102, 108, 111, 118–120, 121, 122, 124, 130*f*
Walton, Rebecca, 13*f*, 25, 95, 100, 110
Western perspectives: academic traditions, 5, 17, 27; collective narratives, 149; cosmovision, 101, 130, 132; role of interpreter and translator, 105; technical communication, 95; translation and interpreting systems, 6, 21, 114; untranslatable terms, 67
whiteness (*blanquitud*), 94–95
Whites, 153*n4*
Wible, Scott, 36*t*, 44–45, 70
Wilson, Shawn, 13*f*, 14, 15, 68

workshops, professionalization, 77, 80*t*, 85, 142
worldview. *See* cosmovision
writing systems, 3–4, 16, 19

Xiang, Zairong, 19, 21
Ximénez, Francisco, 4

Yajima, Yusaku, 24
Young, Iris Marion, 24, 98
Yuto-Nahua linguistic family, 3, 4 13, 14, 19–20, 46*f*, 48*f*, 95, 96*f*, 106, 108, 153*n1*

Zantjer, Rebecca, 26, 153*n6*
Zapoteco linguistic family, 53, 54*t*, 59*t*, 71*f*, 75, 76*t*, 77*t*, 95, 96*f*, 98, 111, 123

ABOUT THE AUTHOR

Nora K. Rivera is an assistant professor in the Department of English at Chapman University. She graduated from the University of Texas at El Paso with a PhD in rhetoric and composition. At this institution, she received the Rhetoric and Composition Outstanding Research student award in 2019 and the 2020–2021 Outstanding Dissertation Award from the College of Liberal Arts. She also holds a Master of Business Administration in marketing and a Master of Arts in Spanish literature and linguistics. Her dissertation, "The Rhetorical Mediator: Understanding Agency in Indigenous Translation and Interpretation through Indigenous Approaches to UX," received the 2022 Outstanding Dissertation award from the American Association of Hispanics in Higher Education (AAHHE) and the 2022 Honorable Mention award from the Latin American Studies Association (LASA).

Rivera's research centers on Latinx and Indigenous rhetorics, Borderland rhetorics, composition, and technical and professional communication. Her multidisciplinary work has been published in the journal *College Composition and Communication*; the *Chicana/Latina Studies Journal of Mujeres Activas en Letras y Cambio* (MALCS); *Programmatic Perspectives*, the journal of the Council for Programs in Technical and Scientific Communication (CPTSC); *Technical Communication*, the journal of the Society for Technical Communication; the *Journal of Teaching Writing*; and *intermezzo*, the online series of *enculturation*. To find out more about her work, visit her website.